MAN IN BIOSPHERE : A CASE STUDY OF SUNDARBAN BIOSPHERE RESERVE

MAN IN BIOSPHERE
A Case Study of Sundarban Biosphere Reserve

Editor
D.B. Mandal

GYAN PUBLISHING HOUSE
New Delhi-110002

MAN IN BIOSPHERE : A CASE STUDY OF SUNDARBAN BIOSPHERE RESERVE

Rs. 540

ISBN : 81-212-0945-5

Published in 2007 in India by
Gyan Publishing House
5, Ansari Road, Darya Ganj, New Delhi-110002
Phones : 23282060, 23261060, Fax : (011) 23285914
E-mail : books@gyanbook . com
Website : http ://www. gyanbooks. com

Laser Type Setting at : Master Computer's, Delhi
Printed at : Salasar Imaging System, Delhi

Contents

Foreword

The present volume is an outcome of research undertaken by the Anthropological Survey of India under its Ten Five Year Plan Project-*Cultural Dimension of Tourism* reoriented as *Man in Biosphere. Sundarban Biosphere Reserve* is one of the Estuarine Biosphere Reserves declared by the Govt. of India in the month of March, 1989 and it is one of the world heritage sites due to its rich biodiversity. Sundarban Biosphere Reserve is spread over to one of the largest mangrove delta of the world i.e. Ganga Brahmaputra Delta. It extends over 9630 sq.kms. in South 24 Parganas district of West Bengal. Sundarban represents an unique biosphere in our country where man nature conflict is increasing day by day due to the population pressure in the lower Gangetic plain of West Bengal. Though in the core area, there is no village but in the buffer area has a total population of 37,55,924 with a density of 930 persons per sq. kms. The fragile ecosystem of Sundarban should be saved if the whole delta is to be saved. Tourism is going to play a dominant role in the coming years because West Bengal Govt. is promoting tourism in Sundarban. The tourist flow to Sundarban is around 2,30,000 in the year 2003-2004 and it includes pilgrims who are going to Gangasagar mela. The scientific personnel of Eastern Region, Kolkata headed by the Core Member, Dr. D.B. Mandal have done an excellent job in Sundarban Biosphere Reserve. The issues which have been covered are tourism as a potential resource for development, impact of tourism on Cultural landscape of Sundarban, structure, cognition and symbolism of forest in the folk tradition, contact and conflict in Sundarban, emerging delta culture, economic condition of the delta villages, human elements in conservation process and evolution of eco tourism in Sundarban.

I am sure this volume is going to be very useful for the academic world as well as to administrators, planners and managers of the forest.

(V.R. Rao)
Director-in-Charge

Preface

The existence of life for last billions of years is the uniqueness of the planet earth and the human being, in the evolutionary process, is probably the best creature on the earth, having the culture with him that leads him towards techno-evolution and help him to extract nature and its components at the utmost level. Air, water, land atmosphere and living organism each of these elements is of vital importance in some form or the other to human kind but uncontrolled and unplanned utilisation of the natural resources cause enormous biodegradation and loss of biodiversity.

Since time immemorial, man has a greater attachment toward forest with inhospitable terrain of mountain, deltaic and any other natural landscape designated now as biosphere reserve. The beauty of nature and diversity in plant, wildlife, physical, culture landscape attracts people of different types to enjoy nature or to know the mystery of it and to feel adventure or to get spiritual joy thus accelerating tourism. With the evolution, growth and development of human civilization tourism gradually turns into a pleasure activity and ultimately with the development of science and technology, tourism industry has evolved into one of the largest and fastest growing industries of the world, having multiple economic benefits. But in the same time, increasing tourism has some adverse impact on biosphere reserve. There is a growing need for acquiring more and more land mass under the natural setting of forest and other lands for formation of new settlement, new industry, and more extraction of natural resources in a very uncontrolled and unplanned way which results in environmental hazard, loss of natural resources, extinction of several plant and animal species, change in climate leading towards global warming and so on. In addition to this local people mostly depend on harvesting the natural resources of the reserve as employment is inadequate. As a result the reserve is also disturbed in some other ways. Hence Biosphere Reserve is an effective conservation unit and a representative part of natural and cultural landscape extending over a large area of terrestrial or coastal/marine ecosystems or a combination thereof. It will serve as a storehouse of information to the researchers who may try their best to find out the way to restore all qualities of the biosphere keeping the area free from damage and pollution as well as to reduce the conflict between man and environment.

With this idea, the concept of Biosphere Reserve was launched by the UNESCO in the year 1970 through Man and Biosphere (MAB) programme. MAB was initiated by UNESCO to develop rational use and conservation of resources of the biosphere. Conservation of landscapes, ecosystem, species and genetic variation will promote economic development, which are ecologically sustainable and are important aspects of Biosphere Reserves. About 482 sites in 102 countries of the world have been demarcated as Biosphere Reserve, of which 14 at present are in India on the basis of their unique biodiversity, natural landscape and effectiveness as a conservation unit.

Sundarban Biosphere Reserve (SBR) is one of them and attains world heritage site in 1989 and represents a rich biodiversity of flora and fauna. Sundarban, one of the largest mangrove deltas in the world, is formed by the great alluvial depositions carried by the three rivers, the Ganga, the Brahamaputra and the Meghna. It extends over an area of 9630 Square kilometres in the state of West Bengal of which 4263 Square kilometres are demarcated as forest area and present a unique situation of biodiversity. It is the only tropical semievergreen mangrove reserve in the world inhabited by the Royal Bengal Tiger. It is the longest biodiversity on globe with 81 plant species and 1586 species of protista (protozoa) and animals. The SBR consists of 102 islands of which 54 are inhabited with different ethnic groups comprising tribes, castes and of different religious faiths like Muslims, Christians. The ecological setup of land, forest, water, wildlife and dependency on the forest in the backdrop of the limitation of agricultural development may be considered to play vital role in forming in belief system-the basis of structure and cognition of the forest.

The people worship deities like Bonobibi, Dakshin Rai, Pir and Gazi Saheb, Manasa, etc., which popularly is known as Bonobibi cult.

The people live in around the reserve and many of them are landless labourers. They live on agriculture, fishing practice of aquaculture, harvesting timber and fire wood, plying boats in general. In fact the local people thrive on the resources of the reserve. Therefore a sustainable management is required to improve the standard of people and check degradation of forest areas. For this effective management, the Anthropological Survey of India has undertaken a National Project on 'Cultural Dimension of Tourism', during the 10th Plan, for systematic investigation and research, which afterwards is reoriented as *'Cultural Dimension of Tourism-The Sundarban Biosphere-Reserve'* in order to understand changes in respect of environment, culture, people's involvement, and interaction, people's idea on sustainable development or eco-development process in contemporary situation in the area of tourism operation, to assess the possibility of growth and development of tourism in the context of population influx and environmental degradation, to evaluate the role of various agents/players in the sustainable development and conservation of Biosphere Reserve. Ten researchers have conducted fieldwork in order to see tourism as potential resource with its limitation, level of cultural landscape in present set up, folk tradition and role of forest with dynamism, contact and conflict, cultural setup in delta region in new situation, economic variability in adjacent village of the reserve, tribal setup and their condition and to assess strategies for ecological balance with its role of management or to evaluate eco-tourism activities in the reserve. These are found pertinent at this stage to conceptualize the SBR holistically. The fieldwork was followed by analysis of data and report writing intervened by a series of workshops to share dialogues between the researchers and the experts from different social sciences and allied disciplines both from the Survey and outside to identify the components and issues on SBR. The present volume is the outcome of this rigorous exercise.

The present research is very much relevant at this era of global warming and biodegradation where people along with flora and fauna are struggling for their specific existences. It is better for us to accept that the prevailing events at SBR would definitely put impact on the developments in the concerned areas. We feel that our contribution will shed light on important areas where the strategies to unravel the untapped potential of the region can be applied and that will probe for further research and help planners and agencies, both government and non-government to implement specific measure for development of the people and the environment in the whole SBR area. All it needs is some determined imaginative entrepreneurs with a long-term vision to exploit it for future benefit. At its various levels we are thankful to many persons and agencies for their guidance, support, help and coordination extended to us without which we cannot do this research and that we duly acknowledge. First and foremost we are thankful to our Director in Charge, Dr. V. R. Rao for his innovative vision, guidance, all sort of encouragement and support and his timely intervention in reorienting the project from mere consideration of tourism development towards the vast area of bio-cultural diversity that aimed towards development of the settlers in Biosphere Reserves. We are also grateful to him for timely publication of this volume. Our sincere regards are due to Prof. S. B. Roy, Chairman, IIBRAD, Kolkata; Dr. J. K. Sarkar, Former Director-in-Charge, Anthropological Survey of India, Kolkata; Dr. M. K. Raha, Former joint Director, Anthropological Survey of India, Kolkata; Prof Kunal Chottopadhyay, Faculty Member of Economic Department, Indian Statistical Institute, Kolkata; Dr. H.S. Debnath, Nodal Officer, SBR, Botanical Survey of India, Kolkata; Shri A.K. Raha, Director, SBR, Salt Lake, Kolkata and Shri M. Sengupta, Director, WBTD, Kolkata. for their kind guidance and suggestions at various stages. Further we are thankful to Dr. A.K. Singh, Senior Human Ecologist, Head Office, Kolkata for his necessary cooperation and guidance extended to implement the project. Lastly we convey our sincere thanks to all administrative and technical staffs of ERC, Kolkata and Head Office, Kolkata for all sort of assistance.

Kolkata
16.01.2007

D. B. Mandal
Core Member, SBR Project

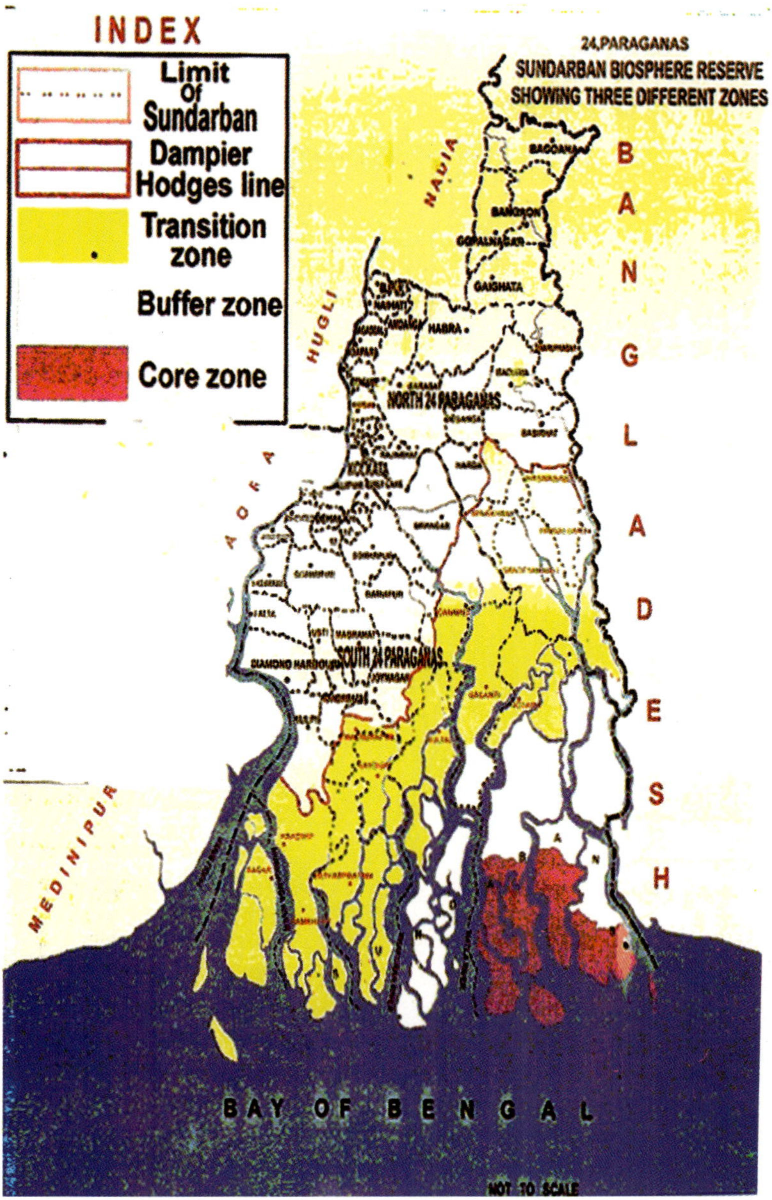

* Boudaries e.g. International, District, Blocks may be shown in the Legend

Introduction

D.B. Mandal

Biosphere Reserve is an effective conservation unit and a representative part of natural and cultural landscape extending over a large area of terrestrial or coastal/marine ecosystems or a combination thereof. It will serve as a storehouse of information to the researchers who may try their best to find out the way to restore all qualities of the biosphere keeping the area free from damage and pollution as well as to reduce the conflict between man and environment. With this idea the concept of Biosphere Reserve was launched by United Nations Educational Scientific and Cultural Organization (UNESCO) in the year 1970 through Man and Biosphere (MAB) programme. MAB was initiated by UNESCO to develop rational use and conservation of resources of the biosphere. Conservation of landscapes, ecosystem, species and genetic variation will promote economic development which are ecologically sustainable and are important aspects of Biosphere Reserves. About 482 sites in 102 countries of the world have been demarcated as Biosphere Reserve.

MAB committee was first constituted in 1972 and was co-ordinated, serviced and funded by the Ministry of Environment and Forests, Government of India, New Delhi. The committee identified 13 sites in India till July 2001 on the basis of their unique biodiversity, natural landscape and effectiveness as a conservation unit. Sundarban Biosphere Reserve (SBR) is one of them constituted in 1989 and represents a rich biodiversity of flora and fauna.

Natural Landscape

Sundarban one of the largest mangrove delta in the world, is formed by the great alluvial depositions carried by the three rivers *viz.*, the Ganga, the Brahamaputra and the Meghna. It is situated between 21°30' and 22°45' north latitudes and 88°06' and 89°05' east longitudes and extends over an area of 9630 sq.kms. in the state of West Bengal of which 4263 sq.kms. is demarcated as forest area and presents a unique situation of biodiversity. It is the only tropical semi evergreen mangrove reserve in the world inhabited by the "Royal Bengal Tiger". This mangrove vegetation stabilizes the shore line, controls erosion and flood, protects the metropolis of Kolkata and the suburbs from the frequent gales originated from Bay of Bengal. So, this Biosphere Reserve plays an important ecological balance.

The region is bounded by the Hoogli River in the west, the Bay of Bengal in the south, the Harinbhanga, Raimangal Rivers along with international boundary with Bangladesh on the east and the Dampier–Hodges line on the north. This line was drawn on the basis of salinity of water. Sundarban Biosphere Reserve consists of 54 inhabited islands out of 102 islands, innumerable rivers, rivulets, creeks and mangrove forests. This area is formed with thirteen blocks of South 24 Parganas and six blocks of North 24 Parganas.

The Sundarban Biosphere Reserve is declared as a "world heritage site" in the year 1989 with its uniqueness as follows (i) the largest mangrove patch of the world, (ii) only mangrove forest inhabited by tigers, (iii) the longest biodiversity on globe with 81 plant species and 1586 species of protista (Protozoa) and animals.

The Biosphere Reserve is divided as core zone with an area of 1330 sq.kms. which is highly protected site for conserving biological diversity and for undertaking non-destructive researches and low impact uses like ecotourism, education, etc. This zone includes a part of the Sundarban Tiger Project and Sundarban National Park (1330 sq.kms.). The Buffer zone covers an area of 1255 sq.kms. of mangrove forest which includes Sajnekhali wildlife sanctuary with other forest blocks. Therefore, reserve forest area of SBR (4263 sq.kms.)includes Sundarban Tiger Reserve (2585 sq.kms.) with Sundarban National Park–core area, Sundarban Tiger Reserve buffer zone and reserve forest (1678 sq.kms.) of south forest division of 24 Parganas. The transition zone covers an area of 5367 sq.kms. which is extended up to boundary of Dampier–Hodges line. This is a dynamic zone of human settlement where agriculture and aquacultures are extensively practiced within local eco-system.

History of Settlement

The name of the forest of Sundarbans was referred to in the *Mahabharata* and the *Puranas*. The story of Chandsagar and the account of Todarmal referred to the existence of a vast swampy forest of this area. History said that Pratapaditya of *Barobhuiyas* ruled over this area towards the end of the16th century. It mentioned that the fort at the south western extremity of Pratapaditya's Kingdom was at Sagar Island (Barun De, 1994: 77). There was a naval station and a ship building yard also. Gangasagar of Sagar Island is regarded as a sacred place related to the legend of meeting of the holy Ganga with the sea to make alive sixty thousand sons of Sagar, King of Ayodhya, with her touch. Thus, except some scattered evidence, record of history of settlement in the area under Sundarban Biosphere Reserve was almost absent till 1770. Forest of Sundarbans was being slowly and steadily cleared and brought under plough and settlements by immigrants from outside with the effort of Claud Russel who gave *ijara* to some people to make the area cultivable clearing the jungle (between 1770-73). During Tillman Henckell, the then, District Magistrate of Jessore, Mr. W. Dampier, Sundarban Commissioner and Lieutenant Hodges, Surveyor took the pain to demarcate the boundary of Sundarbans (1822-30). Henckell distributed the land lot-wise to persons and thus zamindari system was introduced in this area. Thus, expansion of agricultural land was started. It is said that the name of Hingalganj was coined after the name of Henckell. Most of the Zamindars and Talukdars came from the neighbouring Jessore, Khulna and Nadiya districts and northern part of the then 24 Parganas itself. *Pattanidars* and big *rayats* too emigrated from same areas, as also from Kanthi, Tamluk and eastern part of Midnapur Sadar subdivision. The working force comprising the bulk of small peasants, share croppers and landless labourers immigrated into Sundarban region from the drought and famine prone areas of western part of Midnapore Sadar subdivision, Bankura subdivision, Asansol subdivision of Bardhaman district, Singhbhum and Santal Parganas of erstwhile Bihar.

People

Santals, Mundas, Oraons, Kurmis, Koras and persons belonging to depressed Hindu castes came here to clear the jungle. It is noticed that a large number of refugee population came at Sundarbans after seventies. At present out of 102 Islands, there are 2151 settlements are found in only 54 Islands. Out of total population (37, 57,356 as per census, 2001) Scheduled Caste population is 18, 24,955 and Scheduled Tribe population is 1,80,429.

The Brahmin, Goala, Kayastha, Tanti, Mahishya, Napit, are the major castes of this region. The Scheduled Castes living here are the Poundra Kshatriya, Bauri, Bagdi, Chamar, Jele Kaibarta, Namasudra, Malo, Rajbanshi, etc. Poundra Khatriyas dominate the area. The Scheduled Tribes like the Bhumij, Mundas, Oraons, and the Santals inhabit this region. A good number of Muslims and Christian population are also found in the transition part of Sundarban Biosphere region. Settlers coming from different ecosystems, different religions and different communities have developed a unique culture which emerged due to interaction with the environment in this fragile ecosystem for survival. To overcome the natural hazard, people are worshipping the same Gods and Goddesses like Bonobibi, Dakshin Rai, Pir and Gazi Saheb, Manasa, etc. With the above mentioned backdrop of the Sundarbans, the Anthropological Survey of India has undertaken a research project to evaluate the process of Biosphere conservation strategies and tourism activities in this fragile ecosystem with an emphasis to eco-tourism and sustainable development.

Objective of the Study

- To understand the impact of Eco-Tourism as a tool of Eco-Development process.
- To assess the possibility of growth and development of tourism in the context of increasing population and environmental degradation.
- To understand the impact of Eco-Development process as a source of alternative livelihood for the people in the villages of transition zone.

- To assess the contribution of the process of sustainable development towards the conservation of Biosphere Reserve.
- To evaluate the role of various agents/players in the sustainable development and conservation of Biosphere Reserve.
- To understand the level of cultural landscape in the form of land use, economy, lifestyle, peoples' perception, improvement of ethno-tourism in the context of tourism scenario.
- To understand tribal's role in tourism and their living condition.

Thematic Issues

Within the above broad objectives, Anthropological Survey of India has undertaken a National Project on 'Cultural Dimension of Tourism', during the 10th plan, for systematic investigation and research which afterwards is reoriented as '*Cultural Dimension of Tourism: The Sundarban Biosphere Reserve*'. Ten researchers have conducted fieldwork on specific topics at Bakkhali, Gosaba and Gangasagar areas. Gangasagar is known to the people for its religious significance. But Bakkhali and Gosaba have come in light after the arrival of East India Company in general and Andrew Fraser at Fraserganj area and Sir Daniel Hamilton in Gosaba area in particular. The history of settlement in those areas is not more than one hundred and fifty years old.

It is assumed that the reserve is likely to face changes in respect of environment, culture, people's involvement, and intersction, people's idea on sustainable development or eco-development process in contemporary situation in the area of tourism operation. Therefore, tour spots/tour initiating centers/ fringe villages have been selected from the zones of STR or from the vicinity of south forest division of 24 Parganas in order to see tourism as potential resource with its limitation. The level of cultural landscape in present set up, folk tradition and role of forest with dynamism, contact and conflict, cultural set up in delta region in new situation, economic variability in adjacent village of the reserve, are also studied alongwith the tribal set up and their condition and to assess strategies for ecological balance with its role of management as well as evaluating the eco-tourism activities in the reserve.

In the chapter of '*Tourism as a potential resource–a challenging task in the context of contemporary situation of Sundarban Biosphere Reserve*', (S. Gangopadhyay) tourism has been considered as a potential resource in the reserve. Again places of interest on tourism, tour initiating centers have been placed in the zones of STR, Gosaba and reserve forest of south 24 Parganas spreading mainly at Namkhana, Sagar, Patharpratima, Kulpi, Kakdwip and Canning. The ethnic beneficiary groups, i.e., ethno-tourism, by providing local food or local handicraft items, displaying folk dance and drama, momentum of tourism overlooking the environmental degradation and other limitations including land price, contributes in breaking solitude of the forest. The paper highlights the comparitive analysis among the tourism spots spreading in the said zones of SBR.

The chapter on '*Impact of tourism on cultural landscape of Sundarban Biosphere Reserve: A Study at Bakkhali*', (Archana Datta) establishes the impact of tourism industry in changing land use pattern corroborating a new cultural landscape. Changing scenario of cultural landscape of the area is creating new occupational arena and due to interaction with the tourists, the world view of the female folk has changed and they are being interested in participating in the commercial sectors through various occupations. Impact of tourism is also observed in changing cropping pattern. The use of agricultural land of this biosphere has been changed introducing the cash crop like vegetable and chili. Continuation of study among the female folk and discontinuation of study among the male folk due to early involvement in the commercial activities are also another outcome of tourism. Adverse impact of tourism on the natural biosphere is observed in the reduction of natural woodland, accumulation of garbage, polythene bags, etc., ignorance of local people about the sewerage system and selling of soil of agricultural land to the developers for erection of lodges. There is also a marked deterioration in the social sphere due to addiction to alcohol, hemp and gambling. Lastly, the study suggests making a regular bridge over the Hatania, Doania the natural deterrent for easy movement to the tourist centre.

The chapter on '*Structure, cognition and symbolism of forest in the folk tradition of Sundarban*', (Kakali Chakraborty) shows The importance of forest developed under ecological set up of land, forest, water, wildlife and situation of dependency on forest produces. This is studied in the back drop of limitation of agricultural development in oral tradition, rituals, symbols, belief systems and related practices. Thus, folk tradition of Sundarban forms the basis of structure and cognition of forest. It also allows dynamism reflecting the change in role of Bonobibi *i.e.,* from a saviour deity to wealth giving deity.

The Chapter of '*Contact and conflict in the Sundarbans Biosphere Reserve*', (Krishna Basu) has tried to identify nature of contact and conflict of the inhabitants of Sundarban Biosphere Reserve during their procurement both from terrestrial and aquatic resources. Some case studies are projected to establish the contact and conflict situation. It is interesting to note that an effort is made to find out traditional wisdom of the local people during procurement of fish and honey. It is also noticed that due to impact of colonization over a period of time, the area has become over crowded and become a threat to the ecological condition. Introduction of tourism in this area will further aggravate threat to biosphere reserve.

It has been attempted to find out how the culture in this delta region has emerged with the necessity for the survival of the people in the chapter of '*Delta culture of Sundarban: a case study of Gosaba block*' (S.S. Mishra). The people of this region involved in fishing irrespective of castes or communities are identified as a whole as fishing community. Encounter with tigers is a very common phenomenon of this area. The fatal encounter with the tiger during fishing or honey collection make them inclined towards supernatural power in the name of goddess Bonobibi emerged as saviour goddess to all communities irrespective of their castes, creeds and religious affiliation. It is observed that a unique emergence of cultural practice in Sundarbans has evolved with the acceptance and assimilation of goddess Bonobibi as an integral part of traditional Hindu worship. It is also tried to assess the socio-cultural change emerged with the development of Gosaba area under the leadership of Sir Daniel Hamilton.

In the chapter on '*Economic variability of people of Sundarban: an impact study of tourism and Biosphere Reserve*', (Krishna Mandal and Subrata Ray) a switch over from the traditional occupation *i.e.*, agriculture, fishing, honey, wood collection to the service sector such as hotel business, transport including boating, tour guide embroidery work engaged due to gaining popularity in tourism has been traced through an in-depth study at Pakhirala village lying adjacent to STR. The chapter has categorised occupation into traditional occupation mainly, traditional occupation with other occupation and non-practice of traditional occupation. Bagdi follows mostly traditional occupation while Tanti, Kamar and Turi do not follow traditional occupation in most numbers.

The chapter of '*Strategies for ecological balance: management of Sundarban Biosphere Reserve*', (Ratna Dhar) makes an attempt to look into the issues related to the development concerning the people and the conservation initiatives through a village study in the fringe area of Sundarban Tiger Reserve where the eco-development programmes have been initiated. It is observed that the declaration of Biosphere has had a greater impact on the economy and sustenance due to prohibition of resource exploitation from the traditional source of resources. It is found that the people of this area a took the tiger prawn seed collection as alternative way of earning. But this also aggravates river bank erosion for continuous dragging of nets. In addition, use of fine nets for tiger prawn seed collection is destroying other species of fishes. With the restoration of irrigation canals, the people begin to produce *boro* paddy and take initiative to increase pisciculture. But absence of infrastructure for marketing the produces discourages the people. It is mentioned that tourism related occupations are very much seasonal. It is suggested that cold storage, good market and regular repairing of embankment are urgently needed for the development of agricultural produces in this area from subsistence to cash crops. It is observed that the effective co-ordination between scientists with their relevant knowledge and the biosphere managers is largely missing, although the different sectors of the society have been involved in the conservation as well as in village level development.

The chapter on '*Evaluation of eco-tourism activity: A case study of Sundarban Tiger Reserve*', (Amitava Dinda) makes an effort to evaluate eco-tourism activities in the Sundarban Tiger Reserve area, Gosaba and its adjoining areas. Gosaba *bazaar* (market), Pakhirala and Dayapur of Gosaba block are the fringe villages of Sundarban Biosphere Reserve and tour initiating centre to Sundarban Tiger Reserve (STR) and also point of tourist attraction and tourist halt. Over the years a steady flow of both domestic and foreign tourists from all over the continent have visited this place. However, the fruits of economy flourishing on account of tourism are not percolating down to the villagers of Pakhirala, Dayapur, Jamespur, Gosaba market area and its adjoining areas as is evident from their economic condition due to their lesser role in tourism as well as other contributing factors like: ignorance, inaccessibility, poor economy etc. Only recently the local people have begun participating as small scale private entrepreneurs like–hotelieas, travel agents, shop owners etc. It is found that the Forest Department of Government of West Bengal is the key player in organizing eco-tourism and in maintaining balance of fragile ecosystem in Sundarban Tiger Reserve. It is observed that the local authority somehow can not follow the guidelines of Indian National Ecotourism Policy. In the management process of tourism, the interplay of three indicators *viz.*, local community, small scale private entrepreneurs and government determine the degree of management in terms of availability of infrastructure, conservation of ecology, benefit to local community and satisfaction in tourists. In case of STR, which has a fragile eco-system, greater participation of local community along with decisive government intervention in management of tourism is suggested to safeguard the interest of the local community and to protect the ecology. This in turn, will ensure maximum satisfaction to tourists and sustainability to local community. It is also added that some infrastructural facilities can help in improving eco-tourism besides development of local economy. The study also highlights the positive and negative impact of tourism.

1

Tourism as a Potential Resource–A Challenging Task in the Context of Contemporary Situation of Sundarban Biosphere Reserve

Sudhansu Gangopadhyay

Introduction

Tourism may be considered as a substitute economy especially in the area like Sundarban region where other avenues for livelihood are restricted due to its salinity in the sphere of agriculture. The challenges do not only increase the burden on tourism, but also aggravate the contemporary situation in the form of environmental degradation raises a serious threat to which sustain livelihood in a square way. The entire Sundarban Biosphere comprises of two-reserved forests *i.e.,* Sundarban Tiger Reserve and South 24 Parganas forest having an area of 4263 sq.kms. together and non-forest area of 5367 sq.kms. with the total area of Sundarban Biosphere Reserve spreading in 9630 sq.kms. However, there is ambiguity in area at different levels of Sundarban Biosphere Reserve *i.e.,* at total area, or at different break up of SBR (core, buffer and transition zone). It is lying over 19 blocks of which 13 blocks belong to South 24 Parganas and 6 blocks go under North 24 Parganas. It is bounded by the Raimangal, Ichamati, Kalindi, Haribhanga on the Eastern part and Hooghly on the western part while its northern limit and southern boundary is determined by Dampier-Hodges line and Bay of Bengal respectively. According to 2001, census it has population of 37, 559,24 with 18.40 per cent literate and density of population 930 per sq.kms.

Objectives

1. To assess tourism as a potential resource.
2. To note down ethnic situation in terms of imparting socio-economic benefit to the local people and in improvement of ethno-tourism by serving ethnic food, folk-culture or making level of awareness of local people about conservation of Biosphere reserve.
3. To assess the possibility of growth and development of tourism in the context of increasing population and environmental degradation. It is also to see how the pressure of tourism is required to shift from one zone to other for sake of eco-development.

Methodology

Field work is done in two parts (1) February–March, 2005 and (2) February, 2006. Field data, and other information are collected from secondary and primary source.

1. Data and information are taken through canvassing schedules at various establishments at the selected tourist spots.
2. Data and information relating to tourism and its limitations are taken from owners of establishments, villagers, officials, tourists and tour operators.
3. Secondary data specially on land, tourist flow are taken from block level office, land reform and sub-divisional office.
4. Tour initiating centers/tourist spots along with neighbouring forest area have been visited. Different people and officials have been negotiated in order to know the relative position between forest, tourism and fishing and its conflicting situation among these variables.

Tourism and Sundarban Biosphere Reserve (SBR): Unique context and position

The tourism potentiality in SBR is believed to have been enriched by nature's beauty in general and particularly by presence of vast sea shore, meeting place of sea and river, journey along river course, pilgrim heritage especially at Gangasagar, thrilling beauty during boat journey in the dense mangrove forest of Sundarban Tiger Reserve criss-crossed by innumerable creeks and channels. In addition, diverse animal and plant community, large number of birds of different types especially with migrant birds at winter are the main attractions for tourists. The Royal Bengal Tiger along with the deer is the main attraction while crocodile, turtles, and crabs, different fishes including prawn are very much attractive to the visitors.

How SBR formed

The Dampier-Hodges line demarcates the area of SBR. The entire SBR consists of two reserve forests–STR including core and buffer area, South Forest Division and non-forest area (transition zone). The two reserve forests cover the reserve forest of the then Bashirhat and Namkhana range.

(A) Sundarban Tiger Reserve–2585 sq. kms.

(a) Sundarban National Park–1330 sq. kms.–7 forest blocks consists of Matla, Chamta, Chhotahaddy, Goashaba, Gona, Baghmara, Mayadeep form core area or Sundarban National Park.

  (i) Core area without primitive zone–1206 sq. kms and

  (ii) STR core only primitive zone 124 sq. kms.

(b) STR buffer zone 1255 sq. kms. consists of 8 blocks Arbesi Jhilla, Harinbhanga, Khatuajhuri, Nethidhopani, Panchmukhani, Pirkhali, Chandkhali form buffer zone.

(B) South Forest Division (24 Parganas)–1678 sq. kms. It includes the then Namkhana Range of forest.

The total area A & B consist of 4263 sq. kms.

(C) Non Forest Area (Transition Zone)–5367 sq. kms. Here is the ambiguity in area because Dampier-Hodges line does not cover the entire Canning, Joynagar, Mathurapur etc.

The total area of SBR (A+ B+C) 9630 sq. kms.

Communication

The total road including metal and earthen takes 8559.52 kms. out of which 6695 kms. goes to earthen type while metal road records at 1884 kms. The railway network is almost nil except at Kakdwip and recently at Namkhana. The river course is the major way of communication. The jetty plays significant role to maintain communication.

Specific Attraction

Walking through forest part at Bakkhali or visit to Bhagabatpur crocodile rearing center or Kalas or Bijoyara beach– nesting ground of the olive Ridgley turtle, Halley day Island, Lothian Island, Henry island, Sundarkati (lying west of Matla river) all lying in the reserve forest in the south division of 24 parganas may be considered as unique opportunity which encourage tourism.

Beautiful handicraft items which create a great interest among the tourist and market at different tourist spots specially at Bakkhali, Sagar, Gosaba. Even some local food items such as *Khejurgur* (a kind of molasses made from juice of date palm) available at winter make the area popular to the tourist especially at Bakkhali and Gangasagar.

Folk dance or song related to local folk deity such as *Bonobibi* make the area popular to the tourist in the evening or rest time available after finishing journey in the forest especially at Dayapur, Pakhirala village lying in the buffer zone of Sundarban Tiger Reserve. Forest Park at Sajnekhali or mythological sites at Netidhopani of buffer zone of Sundarban Tiger Reserve make the area popular day by day.

Challenges

The SBR faces challenges in growth and development in the event of happening some natural phenomena and socio-economic situations as follows:

(1) Increase salinity and reduction of fresh water

Due to a shift in the fresh water flow from the Hooghly river system into the Padma River since 16th century following tilt toward east, major fresh water rivers running between the two major rivers *i.e.,* Hooghly and Padma, get moribund or activated with silting and sedimentation and shifting courses. It affects smooth navigation and causes man animal conflict arising out of food shortage. River journey over Matla to go to STR via Basanti, Gosaba is seriously affected and is dependent on tide. Ultimately it affects smooth tourism.

(2) Sewage Pollution and Oil spilling

Disposal of sewage water and industrial effluents coming from Kolkata and its adjoining area mix with river water of Sundarban through Hooghly and its tributaries. Oil spilling happens due to increasing numbers of mechanized boats/launches which ply over the rivers of Sundarban as major means of transport or fishing or maintaining tourism.

(3) Desperate collection of shrimps and prawn seed

The more want of prawn in Kolkata and abroad make prawn cultivation popular in water bodies *i.e., bheri* in transition zone of the area. This cultivation has made folk people interested in collection of shrimp and prawn seed desperately in hope of getting in more way with following crude method of netting. As a result of it, the juvenile of a large variety of fin fish (50 species), shell fish (28 species) are also caught in the nets which are destroyed. Ultimately biodiversity is affected with the loss of food chain leading to the shortage of food among the wild life of the reserve.

(4) Cyclonic storm, tidal waves and innundation affecting ecosystem of the reserve

Cyclonic storm, high wind speed, is common phenomena in the reserve. Naturally, average height of high tide is normally high (3.5 meters to 5 meters) and average wind speed is 5 to 10 kms. per hours. It ultimately destroys settlement, crop and enhances salinity.

(5) Reduction of Fishing and Agricultural area

The growth of tourism enhances human activities and consequently in expansion of service related areas with development of parks, recreational arrangement and sewage arrangement. As a result, traditional activity *i.e.,* fishing and agriculture get reduced while providing place for tourism activities. As for example, the area of fishing was much more in the vicinity of Bakkhali seashore. But its area is going to be reduced with expansion of tourism services area which is further aggravated due to land erosion caused by forwarding sea shore or siltation at the mouth of Bakkhali river. Due to this, fishing area for drying is also going to be reduced. Similar problem is also reflected to some extent at Gangasagar.

(6) Breaking solitude

Increasing interest to see forest and wild life in buffer zone or adjoining core area, and eagerness in collection of honey and fish make affect the peace of the area which leads towards breaking solitude and ultimately affects wild life. Sundarban Tiger Reserve of Gosaba block may be considered as an example of such a situation.

(7) Land price

Increasing tourism activities in the tourist spot of Sundarban Biosphere Reserve has also an impact on enhancement of the price of the land lying on road point up to 500 m. and adjoining to the center of the tourism activities. As a result of it few agricultural spots are now used for settlements and leading to change in nature of landscape in the vicinity of tourist spots. In some tourist spots, land already purchased or going to be purchased is lying in waitlist for the growth of lodges.

Bakkhali of Namkhana block, Pakhirala or Dayapur of STR may be cited as example for such case. Again in respect of eco-tourism complex of Gosaba, tourism is going to be spread in some other villages of Bally, Bijoynagar and Lahiripur adjoining to STR which are gaining activation in human activities and also in land price slowly.

Places of Interest

In spite of all limitations, the SBR has its heritage in having number of tourist spots either in buffer area or nearby area of core zone lying in STR and transition zone of SBR. It is also to be added the development of tourism center has been initiated and organized from transition zone of SBR, though few tourist spots are seen to be lying in the buffer or nearby core area of SBR. In this way, tourist spots of SBR have been developed centering round STR lying at Gosaba block, and reserve forest of South 24 parganas spreading mainly at Namkhana, Sagar, Patharpratima, Kulpi, Kakdwip, Canning blocks. The tourist spots are as follows:-

South 24 parganas (forest division).

(A) 1. Bakkhali (Namkhana Block) and surrounding areas (Freserganj, Henry island, Jambudwip,Uttar Bakkhali/Kalistan, Lothian island (Namkhana Block), Estuarine Crocodile Breeding centre, Bhagabatpur (Patharpratima Block).

(B) 1. Gangasagar (Sagar Block) and surrounding areas (Light house of Beguakhali village, Bishalakhi Temple, Dhablat; Shibalaya Temple, Sagor; Marine Biological Research Institute, Bamankhali; Sagor Sports Complex, Harinbari, Ramkrishna Misson,Chayer Gheri.

(C) Others. (1) Gobardhanpur (Patharpratima Block), (2) Kalas (Kulti), (3) Haliday Island (KultaliBlock), (4) Scinarkati (Kultali)

STR Eco-tourism Complex

(D) Gosaba

(1) Sajnekhali, (2) Sudhyanakhali (3) Netidhopani (4) Burirdubry (5) Dobanky (Buffer Zone of SBR)

Tour Initiating Centres

(1) Sajnekhali (Buffer Zone) and (2) Pakhirala (3) Dayapur (4) Gosaba (5) Rangabelia (6) Bally (7) Bijoy nagar (8) Lahiripur (Transition zone of SBR).

There are some other places, which may have potentiality in emerging as tourist spots–Piyali, Canning, Kakdwip and Jharkhali of transition zone of SBR.

How to approach SBR

In order to approach SBR, entry can be made depending on selection of tourist spots. If the spots are of South 24 Parganas (forest division) entry is easier to go via Diamond Harbour. Next entry should be made through Kakdwip, Namkhana, Bakkhali or through Kakdwip, Sagar. In case of trip to STR entry might be taken from Kolkata to Canning by train and then journey can be initiated from Canning/Golabari/Nikarighata for Sonakhali, Basanti, Gosaba in one way by river and in the second way for journey to STR, journey may be initiated by bus for Basanti via Bhangar, Ghatakpur, Sarberia and Basanti, Gosaba Kolkata directly. Therefore in order to have entry to SBR Canning/ Sonakhali may be considered as main way for going to STR while Diamond Harbour, Kakdwip, Namkhana as main way for going to tourist spots lying in and around south 24 parganas (forest division). The distance between Kolkata and Canning, Kolkata and Sonakhali, Kolkata and Namkhana are 40 kms., 90 kms. and 75 kms. respectively.

STUDY OF TOURIST SPOTS

Bakkhali

Name and location

The name of Bakkhali seems to have taken from the name of river Bakkhali though some people have different views about the development of its name. However, Bakkhali tourist spot has developed in the fringe area of Lakshmipur Abad and Amrabati village adjacent to south Bakkhali Reserve Forest.

Tourism as a resource

Emergence as a tourist spot

The establishment of Bakkhali Government Tourist Lodge in 1972 may be considered as a turning point for emergence as tourist spot but it has its root as a point of tourism since 1959 when Bakkhali had its first lodge in the name of Kumudamayee though some other lodges such as Sudhamayee, Samanta, Naraynee etc, are repoted to appear more or less in the same time span in the tourist map of Bakkhali.

ETHNIC BENEFICIARY GROUP AND ESTABLISHMENTS
Numbers of Establishments
(Based on first hand survey 2005)

Sl.No.	*Types of establishment*	*Numbers*	*Owners (ethnic Identity)*
1.	Lodges	20	Mahisya 8, Poundra 2, Doloi-2, co-operatives and Govt. Lodges 3, outsiders-5
2.	Food stalls (for taking meals and dinner)	17	Mahisya 7, Poundra-4 Bagdi 1, others 5.
3.	Tea stalls & stalls for taking light refreshments including sweets	22	Mahisya-12, Poundra-5 others 5.
4.	Vegetables stalls	4	Poundra-1, Mahisya-3
5.	Fish stalls	6	Mahisya-3, Poundra-1 Mohemedan-1 others 1
6.	Sugarcane Juice sellers (van)	6	Mahisya-5, Saha-1
7.	Tender coconut sellers (Van)	4	Mahisya-2, Poundra-2.
8.	Handicraft stalls	7	Mahisya-3, others-4
9.	Rickshaw van	40	Mahisya-21, Poundra-14 Others 5
10.	Grocery and stationery	11	Mahisya-3, Poundra-6 Others 2
11.	Photostudio	1	Poundra-1
12.	Only sweet	1	Poundra-1
13.	STD Booth	4	Mahisya-1, Poundra-3

N.B—Seven other lodges are established at Fraserganj side with more congestion at Bakkhali.

Involvement of Local People

It is also said that in respect of lodge owners, people of Kolkata and surrounding areas come forward prominently, though participation of local people and surrounding villages in this establishment is also encouraging but less than half. But in participation in other establishments specially in running tea, food, vegetable, van rickshaw, involvement of local people specially Amrabati, Lakshmipur Abad has largely increased and tourism may be as a resource for them in addition to agriculture. Mahisya, and Poundra are the main beneficiaries in taking advantage out of tourism when it becomes much more operative in present scenario.

Price of Land

The price of land has become high specially along the line of transport right from the point of Bakkhali Government Lodge near the shore line to the bordering area of Fraserganj. Even years back Rs.1-1½ lakh was fixed as price of land per bigha nearer to road while it has became Rs.3 laks per bigha now. As for example hotel Balaka was built about 25 years back when land was priced at Rs.25,000/-. Due to this situation local people have not been able to such invest huge sums of money in this respect. As a result of this people of Kolkata and surrounding areas become owners of land in order to participate in hotel industry. It is reported that about forty persons are in the lists who purchase land along the communication line in the area of Laxipur-Abad & east, middle, west Amrabati with increasing importance to Bakkhali-Fraserganj tourist centre.

Tourist Flow

The tourist flow at Bakkhali is not yet ascertained as a whole due to paucity of data in this respect at sub divisional level. However, a trend can be apprehended by assessing number of tourists at Bakkhali Government Tourist Lodge, which records 5329 domestic tourists and 45 foreign tourists in the year 2004. In another estimate (based on Tourist Civic Amenity Charge) it is recorded that there is arrival of around 35000 tourists in a year. Normally tourists are in much numbers during October to February. The monthly average would be around 2695. The charge is in effect since 2005.

Sale of local food items

Bakkhali recently earns reputation in having local special food item of date palm juice and date palm molasses during winter. The sale of date palm molasses through tourism has earned a good market seasonally .There are about ten gursals which engage about 50 people. In addition to this, date palm molasses have been prepared by households. About 1300 trees are used for this purpose. A person having 200 trees earns about Rupees 30/40 thousands separately.

Magnitude of Transport Services

The magnitude of transport services may be treated to be an indicator of increasing tourism activities in the present scenario. Now about 24 buses and 5 trekkers are on operative stage.

It is also to be mentioned that rickshaw vans numbering 40 at Bakkhali and 12 at Fraserganj play significant role in transport specially to the tourist who tour in the surrounding area of Bakkhali. The fare varies from Rs. 2/-, 12/-, 15/-, 60/-or 100/- depending on the destination or number of journeys to be performed at a time.

Sale of Handicraft Items

Sale of handicraft items is a great source of income to the people of fringe areas *i.e.,* Laxipur Abad and Amrabati villages. Images of deer, prawn, crab and other animals are prepared and sold at the rate of Rs. 2/-, 4/-or at the Rs. 500/-. So it is an immense scope for earning money. Increasing tourism activity degree may be considered as one of the favourable factors in encouraging the in facilitating its handicraft industry.

Challenging Issues

Bakkhali tourist center has some limitations uninterrupted growth. It is a fact that Namkhana block is a center for raising fishes from neighbouring seas and it has second highest fishing boats of 1427, which are of mechanized and non-mechanized types. Again the blocks had places of sunny area of plain type of land, which were found at the mouth of Bakkhali river and Bay of Bengal, Jambudwip, Fraserganj. But due to becoming shallow in depth at the mouth of Bakkhali River due to siltation, the fish boats would not enter in the interior and therefore it was not possible to get fish dry at the bank of Kalishtan. As a result the fish gets dried at the seashore nearer to Bakkhali tourist center which is also not congenial. In order to get fish dry, more land was prepared at the cost of forest. In addition to this, *hetal* leaf is required to be taken from jungle for getting fish dry in the plain sand land. Thus the forest is effected and smell is not congenial causing pollution to the near by tourist spot. Jambudwip is not allowed for getting fish dry in order to stop forest cutting. Apart from this, Fraserganj sea beach or Laxipur Abad is seriously affected by marine erosion ultimately leading to shortage of area for drying fishes. Therefore, fishing land is gradually approaching towards Bakkhali tourist center, which ultimately hampers the growth of tourist center at Bakkhali. In this way there is now a tussle between tourism and fishing ultimately affecting the prosperity of Bakkhali tourist center.

Inundation of River Flow

The beach of Mousuni is seriously affected due to inundation of river water. In current situation Mousuni Island appears in news in a great way due to such inundation and property loss. Incidentally, Mousani beach may be potential tourist spot of Namkhana block.

Problems and Developments

Shortening of area due to erosion, required area for preservation of forest, and for fishing, demand for more area for development of tourism, soaring price of land at current time appear to have operated in contemporary time making the situations difficult among different spheres.

The bridge over Hatania-Doania, alongwith proper publicity, comprehensive network to connect the tourist places with beautification, sale of local food items make the area more attractive. Thus, tourism may act as source of resource to the people of fringe villages and surrounding areas which have got an impetus to rejuvenate life.

GANGA SAGAR

Name and location

The name of Ganga Sagar is derived probably from the name of Sagar, King of Ayodhya and the river Ganga which ultimately brought back life of 60,000 sons of Sagar by washing ashes in form of divisions of 100 channels. Thus, Ganga became sacred stream and the point of confluence of sea and river is pilgrim spot to the Hindus.

Ganga Sagar lies at the Southern tip of Sagar Block and Island and it lies at a distance of 142 km South of Kolkata.

By road, entry can be made from Kolkata to Kakdwip and Kakdwip to Ganga Sagar via. Kachuberia Ghat from where Gangasagar lies 30 km south of it.

By boat, it is also approachable from Kakdwip or Namkhana to Chemaguri where from anybody can reach Ganga Sagar by bus.It is also approachable from Midnapore through Mayagoalinighat.

List of Lodges

Sl. No.	*Lodge Name*	*Status i.e. Govt./religious Institution Public forum*	*Accommodation Open to Whom*	*Year of Establishment*
1.	Sagar Balaka Paridarshan Kuthi PW (Roads)	Govt.	Only PWD employees	2000
2.	Inspection Bungalow Zila Parishad	Govt.	Open to Govt. and public subject to advance booking.	1984
3.	Irrigation and Waterways Bungalow	Govt.	Open to employee of the concern dept	1975
4.	Youth hostel	Govt.	Open to public and Govt staff	1984
5.	DM old Bungalow(South 24 Pgs.)	Govt.	Open to public and staff of DM old Bungalow	1980
6.	Alipur Recreation Club of DM office		Staff and public	1980
7.	Urmi Mihar Bunglow PHE	Govt.	Mainly staff of PHE and also public byAdvance booking	-
8.	Circuit House	Govt.	Mainly to VIP persons	
9.	Sundaram (S DB)	Govt.	Open to public and staff subject to advance booking	1994
10.	Omkarnath Panthanibash .	Omkamath Ashram	Open to public and devotees	1984
11.	Bharat Sevash ram Sangh	Bharat Sevashram Sangh	Public and devotees	1979
12.	Kapilananda Ashram	Kapilananda Ashram	Devotees	1963
13.	Ganga Saga rBhavan	Kapilmuni Charitable trust public		1999
15.	Larica SagarVihar	Larica Indo Pvt. Ltd. Kolkata	Public	1995
16.	Hoglacottage	Sagor Panchayat Samiti	Public	1994

List of Establishments & Beneficiary Ethnic Groups

Sl. No.	*Stall types*	*Nos.*	*Beneficiary Ethnic Groups*
1.	Food	6	Poundra 3 Mahisya 3
2.	Tea Stall	15	Mahisya Poundra 4 Mohamadan 4 Bagdi 1 Other
3.	Vegetable Stalls	1	Poundra 1
4.	Tea & Rice with STD Booth	8	Kayastha 2 Mahisya 2 Poundra 4.
5.	Only grocery	3	Poundra 2 Mohannaclan 1
6.	Grocery and Stationary with STD Booth	3	Mahisya 2 Brahannin 1
7.	Handicraft Stall	6	Tantubai 1 Poundra 2 Mahisya 3
8.	Tea & Sweet and STD booth	2	Mahisya 2
9.	Saloon	3	Mahisya 3
10.	Tender Coconut seller	14	Poundra 6 Mahisya 8
11.	Sugarcane seller and van rickshaw	3	Mahisya 3
12.	Van Rickshaw	153	40%Mohamedanwth Mahisya Poundra in majority

TOURISM AS A RESOURCE

Emergence as tourist spot

The emergence of Ganga Sagar as a tourist spot is seen as recent phenomena with the establishment of bungalow since 1975, though its reputation as pilgrim center is as old as traced back in the long past. However, its reputation as pilgrim center has started with the establishment of various *ashrams* and *maths* specially after 1975 among which Bharat Sevashram Sangh can be mentioned. The establishment of private lodge like Larica Sagar Vihar and Govt. Lodges like Youth Hostel, Zilaparishad Dak bungalow has accelerated the rate of tourism growth.

Involvement of Local People

Participation of local people *i.e.,* Mahishya, Poundra and even Muslims is worth to mention as beneficiary group. The sellers of tender coconut and sugarcane juice are local people. In local transport, van rickshaw plays a good role and ethnic groups of Mahishya, Poundra, and Mohamedan are the main beneficiaries in this respect. STD booths or saloon appears as service establishments.

Apart from daily need establishment *i.e.,* grocery, stationery, tea or food stall, Ganga Sagar has about permanent 25 stalls with 63 temporary stalls which lie in front of Kapilmuni Temple.

The stall owners are the inhabitants of neighboring villages *i.e.,* Rudranagar, Dhablat, Ariyapur and Sridam, Kalibazar, Bagbazar of nearby areas of Ganga Sagar. But majority are from Sreedam lying in Ganga Sagar. They are mainly Poundra, but few are Tanti, Kayastha, and Brahmin, etc.

Being a pilgrim center, Ganga Sagar has numbers of priests who maintain their livelihood by performing worships relating to marriage, sacred thread and death rituals. The priest consist of two organizations–Janakalyan Brahman Seva Samity (31 members) and Purohit Sangathan (about 27 members) who come from neighbouring villages of Mandirtala, Dhablat, Bankimnagar, Mahisamari, Manshadwip and also Ganga Sagar area.

Role of Transport

In local transport rickshaw van plays a good role in carrying tourist from hotel to seashore or Kapilmuni Ashram and other nearby places of Ganga Sagar in which fare varies from Rs.5/—

20/-. Apart from local transport, there is service of bus, minibus, trekkerplying from Ganga Sagar to Kachuberia Ghat. The total number of vehicles are 43 which have engaged almost 250 persons in this service including driver, conductor and helper who gets Rs.120/-, Rs. 60/-and Rs. 30/-per day in addition to commission money on sale being Rs.11/-for driver and Rs.11/-as commission for conductor and helper jointly. Apart from this there are 80 vehicles of Ambassador, Tata Sumo in all whose fare varies from Rs. 300/-and Rs. 400/-per trip in Island.

Sale of local items

Ganga Sagar also produces date palm juice and mollasses in winter, which has become attractive to the tourists. About 10 gur making workshops *i.e., gursal* have been found in which Muslims play a dominant role. Few other caste groups Mahishya, Karmakar participate in making mollasses. Sagar has numbers of places where river channels or creeks join with sea. These areas are very prospective sites to raise raw material for handicraft items. The area lying between Bankimnagar and Chemaguri may be said as prospective areas for such materials. Among them conch, *Shankh, bi-halve* shell are worth mentioning which are available at the confluence of salt and fresh water. The handicraft items are sold in varying prices from Rs.15/-to 500/-.

Tourist flow and revenue collection

Being a pilgrim center and spot lying at sea shore, tourist flow is high specially during mela. It attracts more than 1,70,000 pilgrims annually based on primary estimate by UNESCO. It is observed that youth hostel has received 8590 tourist during 2003-2004, while the flow is highest during Dec-Feb and least in May and Sept.

Revenue Collection and numbers of Stalls (During Ganga Sagar Mela)

Year	*Stalls Allotted*	*Collection in (Rs)*
2000	1210	954014
2004	1140	743930
2005	1053	608955
2006	1250	979915

(*Source*–B.L.R.O, Sagar Block)

Challenging Issues

Siltation of river bed

The most challenging issues in tourism at Sagar Island is the siltation of river bed lying between lot 8 and Kachuberia ghat. As a result of sedimentation, riverbed along with deposit of sand is seen sometimes above water level during low tide. This increases the time taken to cross-river channel and sometimes it takes about 30 minutes to 45 minutes during low tide. Apart from it, many interior channels were cut off from sea and riverbed becomes shallow to a great extent.

Marine erosion and problem of rehabilitation

Submerged Islands like Ghoramara, Lohachora and marine erosion affected villages like Shibpur, southern part of Dhablat, Beguakhali, Botkhali, Bishaluxmipur displace the people raising the need for rehabilitation of the affected villagers. This situation shortens area for cultivation and fishing which ultimately enters into the area of tourism zone and affect sometimes normal activities of this areas. As for example, the tourist spot of Ganga Sagar developed centering round dag no 2101 having an area of 528 acre and non-rayati area, has been given rehabilitation to about 150 families in about 100 acres of land through *patta* in eastern side.

Encroachment of land for drying fishes

Due to encroachment of land for drying fishes and demand for extension of lodges and Ashrams for tourism, a kind of conflict arises in overlapping zones. Fishing also causes bad smell to the settlement areas. It also causes difficulties during Ganga Sagar Mela.

Problem and development

In fact the tourism potentiality has activated economy and life of people not only for Ganga Sagar but also the Island in totality in spite of its struggling with limitation arising out of marine erosion, sedimentation of river bed, more area for rehabilitation of the effected people of submerged island or eroded area and also area for drying fishes. The lack of infrastructure facilities such as arrangement of electricity, transport bottleneck, sitting arrangement at beach, establishment of eco-tourism centre and boating in brakish water effect tourism.

GOSABA ECO-TOURISM COMPLEX

Name and location

The name of Sundarban Tiger Reserve eco-tourism complex is very much related with the constitution of Sundarban Tiger Reserve in December 1973 taking reserve forest of the then Bashirhat range which is now a part of Sundarban Bio-sphere Reserve.

Sundarban Tiger Reserve Complex of Gosaba is basically a place for eco-tourism where numbers of eco-tourists spots such as Sajnekhali, Sudhyanakhali, Burirdabri, Nethidhopani, Dobaki can be viewed to enjoy forest, rivers and creeks and wild life. Apart from getting taste of eco-tourism, one can have view of Tagore Society and its work at Rangabelia or Gosaba with its heritage place. Sometimes, foreigners may have taste village life wherefrom tour is initiated.

It is located in the southern eastern part of south 24 parganas.

How to reach STR

In order to reach Sundarban Tiger Reserve Complex, one has to reach either at Sajnekhali or Dayapur based tourist lodges or Pakhiralaya or Gosaba/Rangabelia at Bally, Bijoynagar, Lahiripur. Canning or Sonakhali which is to be reached by train 40 kms. or by road for 90 kms. respectively may be said as main entry point from where launch is to be avaited to reach to Sajnekhali or Dayapur or Pakhiralaya. Dhamakhali which is also to be covered by road for 75 kms. may also be said as another point where from launch is to be availed to reach Buridabri of Sundarban Tiger Reserve. In this connection it is to be said that sometimes launch can not ply from Canning during low tide for less or almost no water in river bed and so there is arrangement to get launch at Nikarighata which is 4-5 kms. up from Canning or at Golabari 12 kms. up from Canning.

Significance of STR

This eco-tourism complex holds significance in many ways– presence of unique mangrove tiger land, flow of sizeable tourists including foreign tourists, collection of forest produce, revenue earnings, man-nature conflict in the back drop of close juxtaposition of dense forest and settlement area segregated by innumerable creeks and channels and in the event of deadly attempt for collection of honey, fish mainly in order to maintain livelihood, breaking

LIST OF LODGES (STR ECO-TOURISM COMPLEX)

Place	*Sl. No.*	*Lodge*	*Status of ownership Govt./Public*	*People ownership Local/Neighbour/ Kolkata*	*Year of Estb.*
A Sajnekhali	1	Sajnekhali Govt. Tourist Lodge	Govt.	—	1984
B Dayapur	2	Sundarban Tiger Viewpoint	Public	Kolkata	2004
	3	Sundarban Tiger Camp	Public	Kolkata	2002
	4	Sundarban Env. & Eco-dev society	Public (NGO)	Kolkata	2003
	5	Green house	Public	Gosabs	2003
C Pakhiralar	6	Prarrila Tourist Lodge	Public	Palbrala	2004
	7	Swastic Tourist Lodge	Public	Kolkata	2005
	8	Mangrove Tourist Lodge	Public	Balurghat	2000
	9	Madhuban	Public	Pakhirala	1995
	10	Purbasha Tourist	Public	Palhrala	2005
	11	Krihnakung	Public	Kolkata	1993
	12	AramTourist	Public	Pali	1998
	13	Zilla Parishad	Govt.	Pakhirala	1990
D Gosaba	14	Surya Topa	Public	Gosaba	1997
E Rangabelia	15	Tagore Society	NGO		1975

solitude in increasing degree due to plying of mechanized boat, launched for tourism and forest guards with activities.

The emergence and maturity of STR Complex of Gosaba and involvement of people of area as substitute economy other than dependency on forest can be assessed with the establishment of lodges and status of ownership of lodges. But in order to get a full view of lodges of STR eco-tourism complex five places such as Sajnekhali, Dayapur, Pakhiralar, Gosaba and Rangabelia are to be mentioned.

List of Establishment and Beneficiary Ethnic Groups (Pakhirala)

Sl.No.	*Stall Types*	*Nos.*	*Beneficiary ethnic groups*
1	Grocery	3	Tantubai 2 Poundra 1
2	Grocery and Stationery	1	Poundra 1
3	Tea Stalls	10	Poundra 5 Mahisya 3 Tantubai 2
4	Saloon	1	Napit 1
5	STD WLL Phone	1	Tantubai 1
6	Tailor, Fish, Poultry, Medicine, Cloth	1 each	Poundra-1 in each stall
7	Cement and Sand,	2	Poundra & Tanti

For the purpose of carrying of goods or movement of tourist Pakhiralar has about 35 van owners, which are mostly Poundra or Mahisya and the inhabitants of the village.

Role of boat in STR

Boat movement is very much common and boat is locally called *Bhatbhatti*. This *Bhatbhatti* is inevitable not only for movement of tourists but also for carrying items of daily needs or requirement for lodges. This *Bhatbhatti* plays a significant role in carrying passengers, tourist and cargo and according to a member of executive body of *Bhatbhatti* association formed in 1980 at Gosaba, there are about 150 *Bhatbhatti* of different types *i.e.*, luxury, passenger or cargo *Bhatbhatti* under this association. In addition to passenger boats normally plying in the route, there are also some other boats used for tourism, which are normally taken by travel agency at Canning/Sonakhali or Gosaba. Luxury *Bhatbhatti* are hired by different government departments such as forest, agriculture, health, BDO, etc. In Gosaba *Bhatbhatti* is normally owned/hired by lodges. As for example, Sajnekhali tourist lodge has 10 boats used for tourism under Sajnekhali Mini Boat Association. The boat owners are the inhabitants of Jamespur (5), Dayapur (1), and Satjelia (4) all lying near STR. Similarly Sundarban Tiger Camp a Dayapur based tourist lodge possessed two luxury boats and 2 buses. Pakhirala as an important tour-initiating center has numbers of lodges and *Bhatbhatti* of different uses such as passenger's movement, fishing and tourism. The boats are hired also to any other purpose by different departments. In this way Pakhirala has number of boats as follows:-

LIST OF BOATS

Pakhirala

Hamlet	*Passenger movement*	*Fishing*	*Tourism or any other purpose*
Dakshinpara	3	2	11
Paschimpara	-	-	5
Purbapara	-	-	2
Total	3	2	18

Again it is also said that the boat owners of this village are mostly Poundra and Mahisya. The lodges of Gosaba sometimes hire *Bhatbhatti* or few of them have 1 or 2 boats. Suryatapa lodge uses one boat for tourism. Thus, it is said that in conducting tourism, Pakhirala is one of the leading centers though Jamespur, Satjelia, Dayapur, Gosaba, Bally, Lahiripur also play a leading part in conducting tourism.

Tour agents

There are number of tour agencies who conduct tour in STR. The tour agencies arrange tour package, hotels, food and transport. The tour agencies are based at Canning, Sonakhali, Gosaba and Kolkata. Some of them are Capcika Tour and Travel, Taj Tours and Travels, Wanderlust (Kolkata); Canning Ghosh Travels, Lokenath Travels, Sea View Travels, Ashoka Travels, Sangrila Travels etc. (Canning); Patra Tour and Travels (Kharagpur).

Tourist Flow

According to an estimate based on annual report 2003/2004 of Sundarban Tiger Reserve the tourist flow is increasing.

Tourist Flow

STR Eco-tourism

Year	*Indian*	*Foreign*	*Total*
2002-03	54719	562	55281
2003-04	59225	636	59861

In this connection, the tourist revenue amounted to Rs.1508180/-in the year 2003-04.

Price of land

With increasing flow of tourism and rising revenue, price of land is seen to be higher in the tour initiating centers. Pakhirala may be cited as an example. Near tiger point the price of land has now touched Rs.1 lac per bigha and it was Rs. 20/25 thousand per bigha when tourism was not popular. In addition to this it is said that there is good participation in purchasing land in tour initiating centres. Three more plots have been reported purchased to an amount of 5 bighas @ Rs. 75 thousand to 1,20,000 during 2005-2006. In the interior, Pakhirala village shows trend on high price of land @ Rs 40000/-per bigha.

In Dayapur also, there is a trend of high price of land along the line of lodges facing towards Sajnekhali lodge. Already three plots have been reported purchased in last two years to the extent of 8.5 bigha @Rs 22000/to 35000/-per bigha. The price of land is lower than that of Pakhirala and trend of purchase of land is observed here. Same trend is observed in other tour initiating centres. An amount of land 46 bigha as of Dayapur and Pakhirala is devoted for lodge.

Package tour and tariff rate

In STR eco-tourism complex, package tours are organized in order to see places–Sajnekhali, Sudhyanakhali, Dobanki, Netidhopani, Jhilla, Jharkhali. Tour packages are of one day/one night, its two days/two night three days only. Ordinarily tariff rate varies from 300/-to 450/-for double bed and dormitory bed varies from Rs.15–175/-per bed per day. These are seen normally on Pakhirala based lodges. Food and lodging charge is normally high at Dayapur based lodges where tariff rate varies from Rs.1850–3550/-per head with A.C. charge extra. In Sajnekhali, tariff rate varies from Rs.550-700/-for double bed per day with one major meal and breakfast, while in dormitory, rate is Rs.220/-with meal and breakfast.

Sale of local items

STR raises honey and fish which are also vary popular to tourists. Therefore, these items have also good market among tourists apart from its demands in Kolkata and other areas. According to an estimate based on annual report of STR, it raises 20797 kg. Honey being value of Rs.998256/-in the year 2003–2004. The mean annual catch of fishes is 2500 tons.

STR as a source of livelihood

STR is used not only for tourism but also used for maintaining livelihood through honey collection and fish raising. As a result of it, STR provides a source of income by way of earning through entry fee @ Rs. 15/-, launch Rs.200/-, mechanized boat Rs. 100/-, video camera Rs. 300/-, movie Rs. 1000/-, guide fee Rs. 150/-for Indian and Rs. 200/-for foreigners. In addition to this, boat licensing fee for fish or honey or pass registration fee for honey is taken. In totality, STR acts as source of earnings by way of tourism. Its benefit percolates at various levels right from lodge owners, boat owners to van rickshaw.

Tourism as a beneficiary role

Tourism plays a beneficiary role by way of providing scope for sale of village products to the tourists or by providing some grants by tourists in kinds to the villagers and by way of creating awareness for preservation of wild life as seen in Bally village may be cited as ideal example of symbiotic relation between tourists and villagers.

CHALLENGES

Man Nature conflict

In spite of having prospective potentiality in tourism and revenue collection, STR has various limitations which are also very serious. Among it collection for honey and fish with more demand for metropolis Kolkata or increasing tourism has raised man nature conflict as well as disturbed wild life.

Breaking solitude

Similarly, increasing number of boats including *Bhatbhatti*, launch, patrolling in forest as guard, entry of tourist to greater extent causes pollution in the water by oil spilling or breaks solitude of forest which disturbs wild life. These ultimately lead to man animal conflict by way of injuring or killing the neighbouring villages.

Increasing salinity

There is also threat of inundation of water due to heavy tidal discharge and it ultimately damages earthen dam. As a result of it saline water encroaches in the agricultural land and salinity is increased and due to this situation the fertility of land is decreased.

Infrastructural problem

Shortage of drinking water is a greater problem to this area. Lack of publicity in respect of area and infrastructure is also a problem to the tourists. The ghats are to be developed in Pakhirala and Dayapur side. The bridge over Matla at Canning is badly required while function of Basanti Bridge should be immediately started.

Development

However, in spite of having limitation and arising problem in the context of contemporary situation, tourism acts as a resource not only to people but also to the govt. So it should be developed in the way of eco-tourism perspective.

Comparative study on tourism among different tourist spots

Tourism has developed in SBR centering round STR lying at Gosaba Block and in the vicinity of reserve forest of South 24 parganas. The former one (mainly Sundarban Tiger Reserve) is very close to reserve forest and aspect of tourism lies at the heart of wild life sanctuary and reserve forest, while the latter one (mainly Bakkhali, Gangasagar) is relatively away from forest and the aspect of tourism is mainly the sea, beach, little touch on nearby forest.

It is seen that tourism developed in the vicinity of the reserve forest of South 24 parganas which is at Bakkhali and Gangasagar.

The tour centers are in the line of pleasure and pilgrim tourism basically at the initial level though its area of service in tourism is spreading at present. The tourism at Bakkhali starts to spread with the loss of reputation at Fraserganj due to marine erosion and establishment of Govt. tourist lodge in 1972 though its process of development has accelerated with inception of tourist lodge by people of fringe village as well as neighbouring areas. But initiation at Ganga Sagar tourism starts with the inception of Asram, Math and Mission and later it is accelerated by public forum, Govt. enterprises and little by private entrepreneurs. The response of fringe villages here is lukewarm as contrary to that of Bakkhali.

Again, tourism of Bakkhali is developed partially at forest land initially and later on spreading on rayati land, while at Gangasagar tourism starts at non-rayati areas. On the other hand, tourism at STR as commenced with the Sajnekhali Govt. tourist lodge in 1984, is accelerated by the people of the Kolkata and surrounding place and also by people of villages lying juxtaposed with STR in post 2000 period. The basic characteristic is development of more tour initiating centers at nearby villages of transition areas of STR in recent times in order to serve purpose of touring at STR.

In respect of number of lodges at one place, Bakkhali may be said as larger one with Bakkhali Fraserganj as twin tourist centers having maximum number of lodges of 27. Similarly numbers of establishment are maximum at Bakkhali and Gangasagar comes next. STR being as point of object for tourism has 8 numbers of tour initiating centers. Diversification in construction of lodges and establishments at different tour initiating centers and in having boat and van rickshaw for local transport is seen in STR area. The tourist flow is maximum at Sagar probably due to reputation as pilgrim center followed by STR and Bakkhali.

In respect of ethnic beneficiary group, Mahisya is major ethnic beneficiary group followed by Poundra at Bakkhali while at Ganga Sagar, Poundra and Mahisya are almost equal though Mohamadam are also seen specially driving van rickshaw. At STR, Poundra, Mahisya and Tantubai get benefit almost to an equal extent. The beneficiary fringe villages are seen at Bakkhali and Sagar. They are lying close to tourist spots which are Luxipur

Abad and Amrabati in case of Bakkhali and Suryabinda, Sreedam, Kalibazar, light house in case of Ganga Sagar. But in case of STR fringe villages are not lying close to each other and separated by rivers and creeks. The major beneficiary villages are Pakhirala, Dayapur close to Sajnekhali Govt. tourist lodge while other beneficiary villages are Bally, Bijoynagar, Lahiripur, Gosaba, Rangabelia are little bit away from the said Govt. tourist lodge the main center for entry into STR.

Due to the spread of tourism, Bakkhali is affected by soaring price of land, lying adjacent to road point, the tourism activity spreads along the line of transport upto Fraserganj. There is great rushin the booking of land near the said tourist spot. Almost 40 persons have kept land booked for this purpose along the communication line in fringe villages. Now the price of land has became 3 lakhs per bigha. In case of Ganga Sagar, land price has not become high due to its location in non-rayoti area, though its surrounding villages *i.e.*, Sreedam, Kalibazar, Beguakhali where rayoti land exists to a little extent, are affected by land price. Similarly land price affects the tour initiating villages of Gosaba eco-tourism complex such as Pakhirala, Dayapur, Lahiripur and even at Bally, Bijoynagar etc. Near tiger point of Gosaba eco-tourism complex price of land has touched Rs.1,00,000 per bigha. An amount of land of 46 bigha lying under Dayapur and Pakhirala is devoted for lodge.

In respect of popular saleable items, there is little bit difference in two zones *i.e.*, reserve forest of South 24 parganas and STR Gosaba eco-tourism complex. In formal one, items made of shell or sea product are popular. *Khajurgur* (Molasses of date palm juice) is seasonal saleable products in the vicinity area (where tourism operates) of reserve forest of South 24 parganas. But in case of later one, forest products honey and wax are popular. Folk song and dance centering round Banabibi has now become marketable item in this area.

In nature of challenging issues there is also difference between the zone of reserve forest of South 24 parganas and STR complex of Gosaba. Shortening of fishing area due to encroachment of land by marine erosion and due to non entry in the interior forest area has now become important problem in this area.

Expansion of tourism area and shortening of fishing area has now become a clashing problem in this area. This situation is also true in case of Ganga Sagar also. But in respect of STR complex of Gosaba, man nature conflict along with problem of conservation of forest and wild life, breaking solitude of forest, increasing salinity have now become the major challenges. Lack of infrastructure facilities, increasing salinity, lack of fresh water, poor publicity on the merits of tourism of the area and other facilities are common in this region.

Conclusion

Tourism in the Sundarban Biosphere Reserve had its roots at Bakkhalii, Gangasagar and Sundarban Tiger Reserve since British period but its popularity and degree has gained momentum with opening Government tourist lodge since 1980. Tourism has brought rejuvenation in economic and cultural life of the neighboring villages though people of Kolkata or surrounding areas have quickly responded in possessing hotel business. However, in other spheres, establishments are seen to be benefited by ethnic group of tourist center or near by villages. But there is difference in type of tourism or challenging issues which depending on the location of the spot and context. As for example Bakkhali, Ganga Sagar, Gosaba STR eco-tourism complex, Gosaba belong to nature tourism, pilgrim tourism and eco-tourism respectively. The establishment of lodges/stalls indicates evolutionary phase of growth of tourist centers. The sale of handicraft items or sale of date-palm molasses specially at Bakkhali and Gangasagar has also merits to attract tourist apart from its merits in nature's beauty for both the centers and particularly in having pilgrim heritage at Gangasagar. Similarly, availability of fishes of varied types and crabs may be considered as point of attraction. The eco-tourism situation marked by thrilling sensation of boat journey in river and creeks amidst dense mangrove forest, viewing wild life with special attraction of royal Bengal tiger from watch tower, mythological site at Netidhopani, opportunity of enjoying folk dance/drama of Bonobibi/song or seeing heritage site at Gosaba or Tagore society at Rangabelia has rejuvenated the life and economy of the area by way of drawing more and more tourists to interior island of eco-tourism complex.

In respect of ethnic beneficiary group in possessing hotel business, private entrepreneurs of Kolkata and adjoining area have encouraging role especially at Dayapur based lodges of STR eco-tourism complex and also at Bakkhali. Again role of religious institute is prominent at Ganga Sagar in accommodating tourists/devotees. Government entrepreneurs may be said as deciding role to enhance its popularity at the initial stage which is very much reflected at Bakkhali or Sajnekhali of STR eco-tourism complex. The involvement of local people as means of livelihood is much more reflected at Bakkhali or Pakhirala of STR complex in which people of surrounding/tour initiating village play an important role. The tourism as a beneficiary social role is reflected at Bally, Gosaba Eco-Tourism Complex in terms of providing way for sale of village products or grants to the needy villagers in kinds or maintaining awareness about eco-tourism perspective.

In local transport, necessity for movement of tourists in different areas of tour spots is required and therefore rickshaw vans and boats are of much importance. In the two sectors, by and large local ethnic groups especially of particular tour spot have significant role. In all tour initiating centers such as Bakkhali, Ganga Sagar and Pakhirala of STR eco-tourism, van-rickshaw owners belong to local ethnic groups of Poundra, Mahishya or Mohamedan etc. It is also added that Pakhirala a tour initiating unit of STR eco-tourism complex has almost 18 boats of its own which are used exclusively for tourism and are possessed by people of this village. However, use of boat in tourism is also seen at Sajnekhali Tourist Lodge and these are possessed by local ethnic groups of neighbouring villages.

In respect of involvement in other establishment like food, tea, grocery, stationary or telephone booth, role of people of neighbouring villages or tour initiating villages is much more and it is more or less same in all the tour centers–Bakkhali, Ganga Sagar or STR eco-tourism complex of Gosaba excluding Dayapur and Sajnekhali area where tour is organized exclusively on package basis with all service facilities including transport, food etc.

The Sundarban Biosphere Reserve has gradually picked up its importance in terms of tourism since1980 with inception of government lodges, private lodges or Ashrams. In this respect,

STR eco-tourism complex of Gosaba has gained it by way of attracting 59861 tourists including foreign tourists in the year 2003-04, though Ganga Sagar has its reputation of attracting tourist and has more than 1,70,000 tourist annually with revenue collection Rs. 9,79,915/-in the year 2006 (all time high in the recent times) out of stalls.

In spite of having tourism as potential resource, SBR has challenging issues such as increasing salinity due to breakage of earthen dam, over spilling of saline water during tidal discharge, oil spillage in river water, cyclonic storm, breaking solitude of forest and affecting wild life, mixing of industrial/urban effluents from near by Kolkata metropolis, increasing collection of honey or fish, desperate collection of shrimps and prawn seed. With increasing tourism also these are serious difficulties which pose a real threat to degradation of biosphere reserve. The increasing price of land causes hardship to the common villagers as the land is not easily purchasable by them even though it is little away from tourist spots. Similarly, availablity of fuel wood has become rare to the commoners. Eventhough vegetables do not become cheap but growth of vegetables in kitchen garden or plantation of fuel supplying trees becomes positive effect in some cases at Bakkhali area.

The lack of infrastructural facilities either in absence of bridge specially over Hatania-Doania at Namkhana or bridge over Matla or transport bottleneck between Kachuberia and lot 8 ghat, want of electricity at Gosaba Eco-tourism complex or Ganga Sagar and dearth of concrete ghats at Pakhirala, Dayapur or, Bally, Bijoynagar, beach, beautification at Ganga Sagar, poor publicity of the tourist spot, lack of interest in creation of eco-tourism site by opening deer park specially at Ganga Sagar or boating at brackish water and improvement of nearby tourist centre hamper the emerging tourist scenario.

Nevretheless tourism is a substitute economy not only to the local ethnic group but also people of neighbouring areas or some section of Kolkata and adjoining area. It is also major revenue collecting source to government or other institutes or attached personnel with them. Therefore tourism requires promotion but with elaborate planning and care on minimum damage on Sundarban Biosphere Reserve.

References

Census for W.B 1951, *District Hand Books of 24 parganas* Govt. press Alipur West Bengal.

Chaudhuri, A.B 2004, *Golden triangle of eco-education of tourism*. Regent Park, and Sarkar, Govt. State, Kolkata–40.

Field Director 2003-04, *Annual Report for Sundarban Tiger Reserve* published, by STR, W.B.

Jana, Debaprasad, 2004, *Srikhanda Sundarban (Bengali book)*, Dwip Prakashan, Kolkata.

Roychaudhuri, B and Vyas, P. 2005, *The Sundarbans* published by Rupa Company, 7/16 Daryaganj, New Delhi–110002.

2

Impact of Tourism on Cultural Landscape of Sundarban Biosphere Reserve: A Study at Bakkhali

Archana Datta

The Sundarban Biosphere Reserve, situated in the state of West Bengal, was designated under the National Man and Biosphere Programme (MAB) in 1989. MAB, an ecological Programme was launched by UNESCO (United Nations Educational Scientific and Cultural Organization) in the year 1970 with the aim to develop rational use of land, conservation of the resources of the biosphere and to improve the relationship between man and environment. Biosphere Reserves are the representative parts of natural and cultural landscapes extending over large areas of terrestrial and coastal/marine ecosystems or a combination of both. In the year 2001, MAB committee selected 12 areas of India to be designated as Biosphere Reserves. The Sundarban Biosphere Reserve is one of them. The Sundarban Biosphere Reserve is formed with 13 blocks *viz.*, Sagar, Namkhana, Kakdwip, Patharpratima, Kultali, Mathurapur I and II Jaynagar I and II, Canning I and II, Basanti and Gosaba of South 24 Parganas and six blocks *viz.*, Hingalganj, Hasnabad Haroa, Sandeshkhali I and II and Minakhan of North 24 Parganas. It is bounded by the Hooghli River on the west, the Ichhamati, the Kalindi and the Raimangal rivers on the east, the Dumpier and Hodges line on the north and the Bay of Bengal on the south. The area lies between 21°45' and 22°45' north latitudes and 88°6' and 89°5' east longitudes comprising of 108 islands (of which 54 are marked by settlement sites) and innumerable rivers, rivulets, creeks and mangrove forests. Sundarbans provide one of the rarest opportunities to have an intense feeling of animal love, bird watching, nature love (that may be forest or marine beauty) and being lost in the midst of forest, backwater or open sea.

The total area of this Biosphere Reserve has been divided into three zones (Debnath, 2002:6 and Raha, 2004:16-17):-

1. The Core Zone–with an area of 1692 km^2 (approx.) lies in the eastern part of Sundarban adjoining Bangladesh border. It is a fully conserved area demarcated by Sundarban National Park (1330 km^2 approx.) where nature exists with pristine glory.
2. The Buffer zone–the adjoining area of core zone (2233 km^2 approx.) is demarcated by reserve forests, a portion of Tiger Reserve, Sajnekhali Wild Life Sanctuary, Lothian Wild Life Sanctuary and Haliday Wild Life Sanctuary.
3. The Transition Zone–with an area of 5705 km^2 stretches beyond the boundary of the Buffer Zone, where people extensively depend on agriculture, aquaculture, fishing and forest.

Attractions of Tourists

Tourism is one of the world's largest industries and one of its fastest growing economic sectors (Romila Chawla, 2003: 146). With the decline in the primary sector of economy over the past few years, tourism has become a central part of development activities in the highlands, forests and coast land to give a boost to the economy of the local people as well as the country. The Government of West Bengal is trying to marketise the natural beauty of Sundarbans to the tourists. Tourists are coming here to enjoy wild life, beauty of forests and beauty of creeks.

1. For enjoyment on the wild life and forest beauty, tourists have to go to the south-eastern part of the Matla River. Here floating tourists are mostly found. Some of them also visit the area centering Sajnekhali Government Tourist Lodge–where no other settlement exists except

the Tourist lodge. So, impact of tourism on the cultural landscape is not so prominent in this part.

2. For enjoyment of the beauty of sea, tourists go to Bakkhali, Fraserganj and Sagar Island. Tourism in Bakkhali, Fraserganj and Sagar Island in the southwestern part is altogether different from the previous one. There are settled villages, so, tourism may have a definite impact on the cultural landscape. Ganga Sagar of Sagar Island is famous for pilgrim tourism. Here mass tourism is conducted during the Makar Sankranti (middle of January), when people come to make a religious dip in the mouth of the Ganga. The Anthropological Survey of India has already worked on tourism in this area. For Fraserganj, it may be said that tourists mostly visit Fraserganj. So, to get definite impact of tourism on cultural landscape two fringe villages *viz.*, Lakshimpur Abad and East Amarabati, have been selected for study.

Objective

Carl Sauer (1925: 315-50) said that cultural landscape is fashioned from natural landscape by the people (*i.e.,* culture group). Here culture is the agent, the natural area is the medium and the cultural landscape is the result. With the introduction of tourism industry, the natural as well as the existing cultural landscape (*i.e.,* settlement pattern, house type, sanitation system, sewerage system, source of drinking water, transport network, transport system, land use system, cropping pattern, etc., may be changed depending on the necessities of tourists and inhabited people. To support the industry; hotels, restaurants, different shops, lodges, etc., grow on the natural landscape and change the scenario. So, the objective of the study is:-

1. to highlight the impact of tourism on landscape due to use of local resources like land, man and forest to run tourism industry, as well as on cultural landscape;
2. to know the perception of local people about positive and negative impact of tourism in the area; and
3. to know the perception of local people as well as tourists about sustainable development of tourism in the area.

Methodology

1. Empirical data have been collected through canvassing schedules among the villagers. Impact of tourism on house type, sanitation system, occupations, cropping pattern and use of forest have been highlighted through this survey.
2. Group and individual discussions with the villagers:- This method would help to find out history of tourism, different problems and prospects of tourism as well as impact of tourism on biosphere.
3. Discussion with the owners of lodges, hotels, restaurants, shop keepers, etc:-This discussion also would help to know the year of opening of their commercial centers, owner's origin, accommodation, facility in the lodges, involvement of local people, their process of disposal of waste materials, etc.
4. Discussion with the tourists:-It has highlighted their opinion about the sustainable development of tourism in this area.

Cadastral maps and land use data of R.S. and L R have been collected to show the changing land use pattern of those villages and to demarcate impact of tourism on the landscape. Census records of 1971 to 2001 were collected from the census office and *Gram Panchayat* office to highlight the changes in the demographic profile and occupational pattern.

Study Area

Bakkhali, the second best sea resort of West Bengal located on the extreme southern part of the Ganges delta at Fraserganj (Mecklenburg) Island encircled by Edward's creek, Bakkhali River and Saptamukhi estuary (Barun De, 1994:15) and Fedric Island. The area lies on the confluence of the Ganga with the Bay of Bengal. Here delta formation is still going on. On the shore of the Bay of Bengal shaded by casuarinas plantations and bathed by sweet sun rays, Bakkhali attracts tourists from all over Bengal and beyond to enjoy her soothing, calm and peaceful nature. It is located 130 kms. away from Kolkata, 82 kms. from Diamond Harbour and 25 kms. from Namkhana. The fieldwork was carried on at Lakshmipur Abad and East Amrabati–a hamlet of Amrabati *mouza*, to highlight the impact of tourism on cultural landscape.

Lakshmipur is bounded by Amrabati *mouza* on the north, Fedric Island on the east, which was once separated from Lakshmipur by the Bakkhali River (now there is no trace of this river between the two. Bay of Bengalies on the south and the west of Lakshmipur).

East Amrabati is bounded by Bijoybati in the north, Lakshmipur Abad in the south and Fedric Island in the southeast, Madhyam or Central Amrabati hamlet on the west and Henry's Island in the east. The village was covered by dense forest till sixties. Tigers, deer, and wild bore inhabited in it. Crocodiles were found in the Engine *khal* of Amrabati. The canopy of green forests provided ideal niches for varieties of bird species. A large portion of those villages was covered by *hental (Phoemix paludosa), garan (Ceriops roxburghiana), gengwa (Excaecaria agallocha), golpatta (Nipa fruit cans), kankra (Bruguiera gymnerhiza)*, etc. till seventies. Now all of these species are found within the jurisdiction of Reserve Forest located at Fedric Island and a few within the villages.

The area is located in the south of Tropic of Cancer but temperature is equable due to its proximity to the sea. Average annual maximum temperature is about 35°C. Monsoon starts around mid June and lasts up to third week of September. Rough weather, generally, occurs between the 15th March and the 15th September. Fair weather is observed from the last week of September to the middle of March.

History of Settlement

The Sundarban forest of South Twenty Four Parganas was being slowly and steadily cleared off and brought under cultivation and settlement by the immigrants from outside since 1770. They came in different waves and settled down in different areas of Sundarbans. The process of colonization of Sundarbans has gone on for more than a century and a half, attracting immigrants for beyond the confines of the district (Barun De, 1994:135). But the Fraserganj area of Sundarbans has come to light between the years 1905-1908 with the name of Andrew Fraser, the Lieutenant Governor of the then Bengal (Rabindranath De, 1991:69). Fraserganj is an island–it was known as Mecklenburg Island but the present Fraserganj hamlet was locally known as Narayantala and was named as Fraserganj after the name of Andrew Fraser who fell in love with this village. According to the local people, name of Lakshmipur-the studied village and Edward's creek came after the name of one relative and a daughter of Andrew Fraser respectively. The original settlers of this village came as *jangal kati praja* (tenant to clear the forest). They came mostly from Midnapur and Howrah districts and mainly belonged to Poundra Kshatriya and Mahishya communities.

It was said for Amrabati that Andrew Fraser brought six persons and made them settled at east Amrabati. Their settlement was named as Prajapara. It is assumed that Amrabati was settled earlier than Lakshmipur. The history of this settlement is more than 100 years. Pankaj Mandal one of the resident of East Amrabati said that his grandfather Guruprasad Mandal used to live on the tree making a shelter there before clearing of forest. At present some remnants of previous specimens of forest are found along the eastern boundary of the village.

History of Tourism at Bakkhali

Tourism at Bakkhali includes Amrabati, Lakshmipur Abad and Fedric Island. Early in the fifties attempts were made by the Government of West Bengal to turn Fraserganj into a sea resort. But due to constant erosion of sea beach there, the attempt was finally given up and a new scheme was chalked out to develop a tourist spot at Bakkhali, 2 kms. further south-east of Fraserganj.

Bakkhali was declared as tourist spot in 1968. There were only three log houses (used as rest house) of forest Department at the spot where present Government tourist lodge exists. Forest handed over the houses to the Government of West Bengal.

The seashore of this area is endowed with a wide sea beach. The Forest Department has planted casuarina trees along its length to protect sea beach. Due to its nearness to the confluence of the Ganga the sea beach got enough muddy silt to turn its sand somewhat clayey till the 1990s. But thanks to the nature, Bakkhali sea beach has become much attractive at present gaining a broad sandy sea beach.

The natural beauty of Bakkhali began to attract tourist since seventies. Tourists sometime did not get accommodations in the

Government Tourist Lodge. So, they had to take shelter in the roadside houses of local people. This phenomenon gave the idea to the people like Kalipada Samanta who started Narayani Lodge–the first private tourist lodge of the area in 1974. The local people on their roadside land attached to their residence started Sudhamayee, Kumudamayee and Ma Kali lodges afterwards. Then people from Kolkata and its suburb came to the area and started to open lodges here. Balaka by Anil Pal Chowdhury of Khardah was started in 1977 and then Sahana's by Smritikantha Banerjee of Dakshineswar in 1979. The lodges of local people were *kutcha* houses with mud wall or wall of split bamboo with roof of straw. Balaka Lodge was of wooden structure with 8 rooms only which had been converted into a building with 32 rooms. On the other hand, Sahana was started with 4 rooms. The house was made of brick wall and straw roof and now it is a 2 storied building with 8 rooms. At present about 22 private lodges came up to accommodate tourists besides Government Tourist Lodge, Holiday Home of Sramik Kalyan Parshad, Holiday Home of Samabaya Swastha Unnayan Samastha Ltd., etc.

Infrastructure for Development of Tourism

For introduction of tourism industry in an area certain infrastructures *viz.*, transport system, accommodation system, boarding system and other associated commercial centers are to be developed.

Transport System

Transport sector is an important component of the tourism industry. Previously, there was a bus route from Namkhana to Fraserganj. Part of that road was encroached by sea. The tourism at Bakkhali is actually started after the opening of the metalled road with a width of about 25 feet in 1973 which was earlier an earthen embankment of 4 feet width only. At present the road has been declared as a part of National Highway No. 117 and is connecting Bakkhali with Namkhana covering a distance of 25 kms. Buses started to ply up to Bakkhali from 1980. There were only four buses up to 1986–one bus was owned by Bhuban Mandal a resident of East Amarabati and other three buses belonged to Namkhana Co-operative. As soon as SD 17 (the luxury bus) began to ply through this route, number of buses increased. At present

26 buses are plying between Bakkhali and Namkhana of which two are local buses having route No. 243 started from 1996–97. With the introduction of vessel service over the Hatania Doania river '*Bhutal*' bus service started from 1998 and began to attract more tourists to the Bakkhali area.

Besides buses, there are 150 van rickshaws to carry tourists. Of them 40 vans are owned by the people of Lakshmipur and East Amrabati. Van rickshaws had come in the scenario for about 20 years. At present 4 boats–owners of 2 from Lakshmipur and 2 from West Amrabati are carrying tourists from fishing Harbour to Jambu Dwip. They charge Rs. 30/-from each tourist and Rs.450/-if it is reserved.

Accommodation Facilities for Tourists

Another essential component of infrastructure for development of tourism is the accommodation sector. It has already been mentioned that there are 22 private lodges, two Government accommodations *viz.*, Tourist Lodge and Holiday Home of Sramik Kalyan Parshad, one is made by Samabaya Swastha Samstha Ltd. (co-operative) and one belong to the Fishery Department (Government undertaking). Among the private entrepreneurs, four owners belong to Lakshmipur, four to East Amrabati, one to west Amrabati, nine to Kolkata and its suburb, one to Haripur of Namkhana block, two to Diamond Harbour and one to Kakdwip. Out of 22 private lodges nine occupied the land of Lakshmipur eight of East Amrabati and five of West and central Amrabati. Between the two Govt. accommodations one is located at Lakshmipur and one at Fedric Island. The accommodation system of Samabaya Swasthya Samstha is located at Lakshmipur, and accommodation system of Government undertaking department is located at West Amrabati.

Fooding Arrangements for Tourists

To run a successful tourism industry supply of quality food to the tourists is the concomitant factor quality.

Banasree Hotel–The pioneer in hotel business started in 1972 at Fedric Island to provide quality food to the tourists. But it is noticed that the hotel business, actually started in this area from

Table–1 : Accommodations for Tourists

Sl. No.	Type of Accommodation	No	Owners belong to Lakshmipur	E.Amrabati	Other	Location Lakshmipur	E.Amrabati	Fedric Island	Others
1.	Pvt. Lodges	22	4	4	14	9	8	-	5
*2.	Govt. Lodges	2	-	-	-	1	-	1	-
*3.	Co-operative	1	-	-	-	1	-	-	-
*4.	Govt. Undertaking	1	-	-	-	-	-	-	1
	Total	**26**	**4**	**4**	**14**	**11**	**8**	**1**	**6**

* **N.B.** * Sl.No. 2 belong to Govt.,
* Sl.No. 3 belong to Samabay Swastha Samstha
* Sl.No. 4 belong to Government undertaking department.

Table–2 : Fooding Arrangements for Tourists

Sl. No	Type	No.	Owners belong to Lakshmipur	E. Amrabati	Others	Location Lakshmipur	E. Amrabati	Federick Island
1.	Hotels	19	2	15	2	-	4	15
2.	Hotel + Tea Stall	2	1	-	1	1	1	-
3.	Hotel + restaurant	1	1	-	-	-	1	-
4.	Lodge + Hotel	1	1	-	-	1	-	-
5.	Restaurant	1	-	1	-	-	1	-

Source: Field survey.

1984 onwards. Now about 19 hotels, 2 hotel cum tea stalls, 1 hotel cum restaurant, 1 lodge cum hotel and one only restaurant have occupied the landscape (Table–2). Out of 19 hotels, 15 are located at Fedric Island and 4 hotels are at East Amrabati. Out of two hotels cum tea stalls one is located at East Amrabati and one at Lakshmipur. One hotel cum restaurant and one only restaurant supply fast food to the tourists are found at East Amrabati, whereas one lodge cum hotel has been established at Lakshmipur. It is observed that people of East Amrabati are more involved in hotel business.

Other Infrastructural Facilities

With the development of hotels and lodges; grocery shops, stationary shops, tea stalls, vegetable shops, poultry, meat shops, sweet-shops have come up. Date molasses (*khajur gur*) and different showpieces made of shell (*jhinuk*) have become very popular to the tourists. Both the business need to involve family members of shopkeepers. Mostly husband and wife do the needful to carry these household industries. There are about ten *jhinuk* shops at Bakkhali except one at Lakshmipur. All have occupied the land of Fedric Island. Seven persons of Lakshmipur, two from East Amrabati and one from out side run this business. At present, due to high demand of date molasses, to the tourists about three person of East Amrabati remain very busy from December to the first week of February to prepare molasses and sell it to the tourists. They make temporary road-side make shift huts at Lakshmipur and Amrabati and live there with their families to prepare it and sell molasses. To supply molasses to the tourists, plastic containers are sold by them as well as by the stationary shop owners. Lodges, hotels, restaurants, tea stall, sweet shops, meat shops, selling of molasses and shell materials are directly related to tourism. Besides these, grocery shops, stationary shops, vegetable shops, poultry, meat shop, fish business, etc., are indirectly related to the tourists through hoteliers. Business of building materials and hardware has come up in this area to supply raw materials to the lodge owners to supplement their necessity in emergency, whereas, at present, many local people become interested to make concrete building. They are also enjoying the advantage of those trades.

For the development of tourism, electricity, drinking water and telephone connection are also an important criteria besides road, lodging and boarding systems. The study area got electric connection by the end of seventies. Lakshmipur got 17 tube wells and East Amrabati got 16–including tube wells owned by the lodges. The area has been connected with the telephone exchange at Namkhana not more than 5 years back. At present 4 telephone booths have been established to serve. The need of the tourists. Among them three belong to the inhabitant of East Amrabati and one to Lakshmipur. Thus, with the development of tourism industry these infrastructural facilities have been developed to meet the demand of the tourists and the local people also are enjoying the benefit of These facilities. Many people of these villages have electric connection and T.V. set and at least two households of each village are enjoying the facility of having telephone. It may be said that with the establishment of those lodges, hotels, shops, telephone booths, etc., a new cultural landscape has emerged in the study area.

Impact of Tourism on Cultural Landscape

To know the impact of tourism on cultural landscape, the changes in land use pattern have to be studied first because development of tourism has direct impact on landscape.

Impact of tourism on landscape

For the present study total land use pattern has been classified under twelve categories *viz.*, (i) *sikosti* (area occupied by sea), (ii) *baluchar* (area occupied by sand), (iii) settlement sites include houses, *bhiti* (abandoned homestead land), graveyard, school and playground (iv) agricultural land (v) water bodies include pond, *doba* (small pond), *nullah* (narrow drainage channel) and canal, (vi) natural woodland (vii) embankment, (viii) man made woodland/ garden, (ix) commercial and industrial area, (x) prospective commercial area, (xi) communication (road/path), (xii) fallow land.

Table–3 : Landuse of Lakshmipur (As percentage to total area)

Sl.No.	Categories of land 1954-57	Area 1954-57	Area Present	Increase or decrease
(i)	Shikosti	—	51.82	+51.82
(ii)	Baluchar	33.60	5.18	-28.42
(iii)	Settlement sites	2.51	4.91	+2.40
(iv)	Agricultural land	53.77	17.90	-35.87
(v)	Water bodies	3.58	4.58	+1.00
(vi)	Natural woodland	3.06	3.21	+0.15
(vii)	Embankment	1.48	0.70	-0.78
(viii)	Man made woodland	0.98	1.52	+0.54
(ix)	Commercial and industrial area	—	5.44	+5.44
(x)	Prospective commercial area	—	3.49	+3.49
(xi)	Communication	0.91	1.10	+0.19
(xii)	Fallow land	0.11	0.15	+0.04

Source:-Settlement office and field survey.

R S map (Map 1) shows that major land was occupied by agricultural use (53.77 per cent land) followed by *baluchar* (33.60 per cent), water bodies (3.58 per cent), natural woodland (3.06 per cent), settlement sites (2.51 per cent), embankment (1.48 per cent), man made woodland (0.98 per cent), road and path (0.91 per cent) and fallow land (0.11 per cent) respectively (table–3). Presence of vast *baluchar* indicates previous encroachment record of sea. The landscape was occupied by unmetalled road and amorphous dispersed type of settlements.

The present land use map of the village shows a different picture (Map 2). It denotes that large areas of the village have been encroached by the sea (51.82 per cent) which was not in the scenario of R.S map. Decrease is observed in the *baluchar* (28 per cent), agricultural land (-35.87 per cent) and embankment (-0.78 per cent). The *Baluchar* have mostly been occupied by sea. Previous natural and man made woodland also have been encroached by sea. For the erosion of forest land a large area of agricultural land have been opened to be covered by sand and salt and become unworthy for cultivation. Now natural woodlands have grown in the western and south eastern part of the village on the agricultural land. To protect agricultural land and houses from salinity, local people also have planted casuarinas trees to obstruct wind. Encroachment of sea is also evident in the settlement sites, canals and embankments (erected to protect the village from encroachment of sea). It is observed that two factors are responsible for the changes in the land use of Lakshmipur. One is the natural and other is the human interference.

Human interference on the landscape of Lakshmipur for the development of tourism has been reflected by the addition of area under industrial and commercial uses (+5.44 per cent). These areas include lodges, hotels, different shops, restaurants, tea stalls, STD booths, petrol pump and other commercial uses. Albeit, one wind energy farm center has been opened here to generate electricity (*Haoa Bidyut Kendra*) and it has shared a large portion of the area under the category of commercial and industrial uses (about 73 per cent of this category). The area under prospective of commercial uses also reflects the impact of tourism. Many people, especially, from Kolkata have purchased those lands for commercial uses in future. These categories of land were introduced mostly through encroachment into the agricultural land (+3.49 per cent) emerging a new cultural landscape in this area. Increasing area under water bodies (+1.00 per cent) may be due to addition of *nayanjuli* which came into the scenario during the construction of road. The soils from this roadside agricultural land were used to raise the platform of the road. As a result these roadside ditches (*nayanjuli*) were formed. Now, this is also used as sewerage channel. It is observed during field work that soil of two agricultural plots were being carried to Dolphin tourist lodge for its expansion turning those plots into ponds. Thus, those very uses of land have changed the character of agricultural plots and raised the area under water bodies. To accommodate the growing population, area under settlement sites has increased (2.40 per cent) along the road side land or embankments forming a linear settlement pattern.

Table-4 : Landuse of East Amrabati
(As percentage to total area)

Sl.No.	*Categories of land*	*Area 1954-57*	*Area Present*	*Increase or decrease*
(i)	Shikosti	—	—	—
(ii)	Baluchar	—	—	—
(iii)	Settlement sites	3.36	8.09	+ 4.73
(iv)	Agricultural land	80.62	77.26	-3.36
(v)	Water bodies	9.47	8.86	-0.61
(vi)	Natural woodland	4.07	0.61	-3.46
(vii)	Embankment	1.15	1.30	+ 0.15
(viii)	Man made woodland	0.12	0.13	+ 0.01
(ix)	Commercial and industrial area	—	0.80	+ 0.80
(x)	Prospective commercial area	—	—	—
(xi)	Communication	0.28	0.97	+ 0.69
(xii)	Fallow	0.93	1.98	+ 1.05

Source:-Settlement office and field survey.

Land use pattern of East Amrabati is somewhat different than that of Lakshmipur. Absence of *shikosti* and *baluchar* depicts the stability of the landscape of this hamlet. R.S map of East Amrabati (Map 3) depicts dominance of agricultural land (80.62 per cent), followed by *gheri* (agricultural land used for fishing), ponds, *doba*, canal and *nullah* (9.47 per cent) natural woodland (4.07 per cent) and settlement sites (3.36 per cent).

In the present land use map (figure-4), area under settlement sites has been increased mostly at the cost of agricultural land (+ 4.73 per cent). Decrease is observed in the agricultural land (-3.36 per cent) water bodies (-0.61 per cent) and natural woodland (-3.46 per cent). Impact of tourism on the landscape is comparatively lower than that of Lakshmipur. Only 0.80 per cent land of this hamlet has been involved in tourism industry.

The Table-3 shows that about 65.07 per cent of land of Lakshmipur has been involved in change, which is only 7.43 per cent for East Amrabati (Table–4). It may be said that about 9 per cent land of Lakshmipur and only 0.80 per cent land of East Amrabati have been involved in change with the introduction of tourism industry. It is note worthythat prospective commercial area is absent at East Amrabati.

Impact of Tourism on Cultural Landscape

Cultural landscape of this area involves the land used for agriculture, communication, settlement and commercial purposes. Communication, and commercial systems of the area have already been discussed earlier. Settlement and agriculture will be discussed ahead.

Settlement

It is found that with the development of tourism industry roadside residential as well as commercial houses have emerged in the form of a linear type of settlement. Albeit, in general, the villages have an amorphous dispersed settlement character, yet along the side of embankments linear settlements are not uncommon.

It is interesting to note that with increasing attraction of earning from tourism industry dual residential system has developed and this system is mostly found among the hotel owners. For running the hotel business, hoteliers mostly engage their wives as cook and their children above 10 years also help their parents in different ways. At the initial stage they used to come early in the morning and go back to their residence late at night and gradually they have established separate residence within their hotel keeping in touch with their original home. Some families stay at the tourist place during the peak tourist season and go back to their village home during lean season. October to February is considered as peak season for the tourists. On the 25th December and the 26th January, pressure of tourists becomes extreme. The researcher was present at Bakkhali on the 25th December 2004 and it was observed by her that Bakkhali tourism reached its saturation point on that day. Many people did not get accommodation till late night and at last they had to take shelter in the houses of the local people on high rent. The author knows one such family who had to take shelter in a room and a verandah of a family against Rs. 500/-.

Houses:-Since, settlement comprises a group of houses, so house type, use of house building materials and sanitary system are important aspects of cultural landscape. Impact of tourism is noticed in the house type also. Previously, the houses were made mostly of mud wall/split bamboo or *hogla* with roof thatched by straw or *bakra* leaf (leaf of *Phoenix paludosa*). But with the opening of road and construction of buildings for lodges the village people aspired to use tiles initially and to protect their roof from cyclone. Many of them constructed concrete buildings (but it has been done only by affluent families). Out of 180-studied houses of Lakshmipur 5 per cent are *pucca* houses, roof of 40.55 per cent houses are made of tiles and 23.89 per cent houses are made with brick wall. Whereas 363 houses of East Amrabati about 6 per cent are *pucca* houses, roof of 31.68 per cent houses are made of tiles and 17.08 per cent houses are made with brick wall. The people of East Amrabati prefer to use straw for the roof of the main house to keep their house cool. Most of the people use tiles for the roof of their verandahs.

Sanitary system:-Previously the villagers had no concept of latrine and bathroom. They were habituated to perform their nature's call in the open air. Now, about 14.44 per cent houses of Lakshmipur have bathroom and latrine both, 1.11 per cent houses have attached bathroom and latrine and 3.33 per cent houses have only latrine.

When East Amrabati is considered it is found that about 14.05 per cent households have both bathroom and latrine and 10.47 per cent households have latrine only. No houses of East Amrabati have attached bath. Higher percentage of latrine at East Amrabati than that of Lakshmipur also proves that more people of this village can afford to spend money for this change. Villagers said that with the clearance of jungle for the development of tourism industry they are facing difficulties and are feeling the necessity to construct bathroom and latrine.

Drinking water:-Source of drinking water is another important infrastructure for the development of tourism as well as the settlement of an area. Earlier people had to use pond water for drinking purposes. They had to mix lime to make the water potable by minimizing the salinity. Now, with the establishment of tourist lodges, tube wells had to be suck to supply potable water to the tourists. Side by side, gram panchayat also sunk tube wells. Before seventies, there was only one tube well belonging to the Prabartak Asram. Now, Lakshmipur has 17 tube wells and East Amrabati has 16 tube wells. People have been habituated to use tube well water for drinking purposes.

Agriculture:-Impact of tourism is also noticed in the cropping pattern of the village in a very low profile and in the change of character of agricultural land. The people of the area used to grow only a single crop (*aman* paddy) in their agricultural field. With the development of transportation system and introduction of irrigation system (using pond water by pump) people become interested to produce chilly and vegetables in their agricultural field. They sell them mostly to Jetty (*Ghat*) and Bakkhali markets.

Fifty two households who have land at Lakshmipur were interviewed to know their preference for selection of crops. Out of them 32 are owner cultivators, 9 are owner cum lease holders, 3 are owner cum share croppers, 6 are only leaseholders and 2 are only share croppers.

Table–5 : Land used for cash crops

Villages	*Own land*		*Leased land*		*Shared land*	
	Chilly	*Vegetables*	*Chilly*	*Vegetables*	*Chilly*	*Vegetables*
Lakshmipur	22.14	7.85	58.39	13.04	42.11	1.75
East Amrabati	8.27	0.74	3.87	0.42	—	—

Source: Field survey.

It is observed that cultivators are producing vegetables like potato, onion, brinjal, turnip, beet, spinach, carrot, etc. in 7.85 per cent of their agricultural land in *rabi* season, whereas it is 22.14 per cent for chilly. The lease holders generally take land from others on lease for *rabi* season to produce mostly chilly.

When East Amrabati was considered it was found that 225 households out of 399 (56.39 per cent) cultivate their agricultural land at East Amrabati. Among them 149 are owner cultivators, 24 are owner cum lease holders, 15 are owner cum share croppers, 1 is owner cum share cropper as well as lease holder, 10 are only share croppers and 2 are lease holders cum share croppers. The table shows that cultivators of East Amrabati are not as interested to produce vegetable and chilly in their agricultural field as preferred by the cultivators of Lakshmipur. Field record shows they are interested to produce *aman* paddy and a few of them produce *boro* paddy also. Some of them can sell paddy to the hotels and villagers. It may be said that as East Amrabati had a stable condition since the inception of its settlement, the cultivators had more land under their ownership and they get their sustenance from the traditional crop, they do not strive to produce any cash crop which is labour intensive.

Impact of Tourism on Occupation Pattern

Occupation pattern is very much related to the cultural landscape emerged due to the development of tourism industry. Because of the business centers grown or business activities or other tourism related activities, they have created sources of earning for the people with the introduction of tourism. The most visible sign in this occupational sector is transfer of labour force from primary to secondary and tertiary sectors of occupations. As comparable data are available from 1981 (census data), to show the changes in occupation pattern, that year may be taken as the base year. Amrabati mouza is considered. The Tables 6 and 7 show that a large portion of population was dependent on primary sector of occupations *viz.*, agriculture and fishing. About 76.81 per cent workers of Lakshmipur and 55.92 per cent workers of East Amrabati and 23.19 per cent and 44.07 per cent for Lakshmipur and East Amrabati respectively were engaged in secondary and tertiary sectors of occupations during 1981.

But dramatic changes are observed during 2001. The total scenario become reversed. Engagement of 74.79 per cent and 80.90 per cent of workers of Lakshmipur and East Amrabati respectively to the secondary and tertiary sectors of occupations in 2001 indicates the impact of tourism industry on occupation structure. Another noticeable fact is the increasing involvement of female folk as earning members of their families. It was only 1.41 per cent and 0.39 per cent for East Amrabati during 1981 and became 6.82 and 3.57 per cent respectively in 2001. Increase of participation of female workers in secondary and tertiary sectors of occupation is more at Lakshmipur than that at East Amrabati. Uncertainty in agriculture and fishing diverted people to different sectors of commercial activities related to tourism industry. With the development of tourism, world view of female folk has changed and they are trying to be regular earning members of their families taking active role to run hotels as cook, to run tea stall, stationery and grocery shops, to prepare date molasses and make shell materials. Sometimes they work as maid servants in the lodges and run tailoring shops also.

It is observed that dependency of the working population of East Amrabati on primary sector of occupations was lower than Lakshmipur since 1981. As East Amrabati was settled earlier than Lakshmipur, pressure of population on agricultural land of East Amrabati was higher than Lakshmipur since 1951(599/km^2) at East Amrabati and only 230/km^2) at Lakshmipur). It may be said that as people of East Amrabati had more roadside land lucrative for the development of tourism industry they could utilize their landed property for the promotion of tourism since the inception of this industry and more people were involved in the tourism based occupations since that time.

The tourism related occupations may be divided into three categories–(i) commercial–it includes different type of businesses *viz.*, lodges, hotels, shops, selling of molasses, keeping of poultry birds, selling of eggs, etc. (ii) services, *viz.*, services at lodges and hotels, banks, etc. and (iii) other work, *viz.*, work as mason, or as assistant of mason (*jogare*), van rickshaw puller, maid servants at lodges, etc. It is observed that the category of "other work" has engaged more male population followed by commercial and service categories, whereas, commercial activities have attracted more female folk followed by 'other' and 'service' categories respectively (Table-8).

At East Amrabati about 60.47 per cent of population engaged in tourism related occupations are involved in commercial activities especially in business of hotel, lodge grocery shop and tea stall and it is much higher than that of Lakshmipur. On the other hand, 'service' category has absorbed more workers of Lakshmipur.

Table–8 : Workers of Lakshmipur involved in Tourism Industry
(Percentage to the total workers engaged in tourism)

Sl.No.	*Categories of workers*	*Male*	*Percentage*	*Female*	*Percentage*	*Total*	*Percentage*
1.	Commercial	41	33.61	22	18.03	63	51.64
2.	Services	6	4.92	2	1.64	8	6.56
3.	Others	43	35.24	8	6.56	51	41.80
	Total	90	73.77	32	26.23	122	100

Source:-Field survey

Table–9 : Workers of East Amrabati involved in Tourism Industry (Percentage to the total workers engaged in tourism)

Sl.No.	*Categories of workers*	*Male*	*Percentage*	*Female*	*Percentage*	*Total*	*Percentage*
1.	Commercial	86	33.99	67	26.48	153	60.47
2.	Services	19	7.51	5	1.98	24	9.49
3.	Others	59	23.32	17	6.72	76	30.04
	Total	164	64.82	89	35.18	253	100

Source:-Field survey

Impact of Tourism on Biosphere and Perception of People

The record shows that about 25.29 hectares of land of Lakshmipur was under natural woodland and it has been reduced to 3.04 hectares to fulfill the demand of tourism industry on the one hand and encroachment of sea on the other. It may be said that more woodland was encroached by tourism industry in the Fedric Island which is an unsurveyed area and belong to the Forest Department. The author had a chance to go to Bakkhali earlier in 1984 and when she revisited the same place in 2004-2005 during fieldwork she noticed that the jungle on the roadside land in front of the Government Tourist Lodge had disappeared. Hotels and shops occupied the road side land. Woods of jungles were used as fuel by the villagers for their own consumption as well as by hoteliers.

People of the village mostly depend on forest for their fuel. Although the Forest Department has restricted the entrance of villagers into the forest, they are allowed to collect casuarina leaves on Sunday only. But it is noticed that villagers, especially, female folk enter the forest at dawn for most of the days, because, many of them earn money by selling fuel to the hotel owners. It is observed that about 24 per cent families of the village do not depend on forest for their fuel consumption. They have planted casuarina and other trees as well as domesticated bovine animals and use cow dung cake, own firewood and straw for cooking, two households are using LPG gas.

The hotel owners of Sahana Lodge said that he used to see deer early in the morning in front of his lodge till the bus route was open. The villagers are loosing the natural beauty with the introduction of transportation system.

During December and January, especially, 25th December, 23rd and 26th January and Saturdays and Sundays of those months, forest of Bakkhali (near the sea coast) are occupied mostly by picnic parties. It was estimated that on the 25th December, 2004 about 5000 people (data supplied by the parking committee who collect parking fees from the picnic parties and others coming by buses or private cars or vans) gathered in that area for picnic or to visit the sea. It was observed on the 26th December morning the sea beach and forest was littered with polythene bags and other waste materials left by picnic parties.

Another problem emerged with the introduction of tourism industry in this area was of improper sewerage system. Whenever pressure of tourists increased in the area the local people suffered due to bad smell in the evening.

Perception of Local People

The local people expressed some positive and negative impacts of tourism.

A. Positive impacts of tourism:-

(1) The development of tourism industry in this area has enhanced the economy of the local people who possessed land.

(2) Value of land of this area has increased due to introduction of tourism. Now, road side land is sold @ Rs. 5 lakhs/*bigha* which was only Rs.1600/*bigha* 25 years ago. Land inside the village is sold @ Rs. 2-3 lakhs/*bigha* which had no demand earlier.

(3) Small fishermen are earning money by selling fish in the Bakkhali market.

(4) People are earning money through poultry. Lakshmipur has 2 poultries. Village people are also becoming interested in keeping poultry birds to sell eggs to hotels.

(5) Earlier, no farmers used to produce vegetables in their agricultural field. Now, they are getting ready market at Bakkhali and Jetty Ghat. So, many of them are producing vegetables.

(6) People, are earning money from van rickshaw and mechanical boat to take tourists to Jambu Dwip, or as bus drivers.

(7) Young people are getting jobs at lodge and hotels.

(8) Many people of the villages are earning money from tea stalls, grocery shops, stationery shops, STD booths, hotels, etc.

B. Negative impact of tourism:-

(1) All the people of the village complain that due to introduction of tourism industry, habit of taking alcohol has increased. Now, more people, especially, the young generation are taking alcohol. According to them most of the shops sell wine, local drinks and hemp (*ganja*) illegally. Many young people are taking alcohol and misbehave with their parents. The guardians are becoming helpless in controlling their children.

(2) More people are being involved in gambling.

(3) Young generations are spending more time in the market place by gossiping with friends. As a result their education is being hampered.

(4) They are preferring to take fast food from the market and thus making misuse of money.

(5) The village people are missing to see the wild animals, especially, deer in their locality.

(6) Forest is loosing its density due to illegal selling of forest wood to hoteliers.

(7) Some hotels are being used for sex business creating an unhealthy atmosphere for the society.

(8) Villagers have no place for disposal of solid wastes of hotels and lodges.

(9) There are no proper sewerage systems in the area.

(10) Most of all the people of Lakshmipur are at their wits end with the news of Sahara (a big business company) coming to their village for promotion of tourism industry. Many people apprehend that they cannot renovate their houses or cannot make any plan for

cultivation. Even they cannot cultivate fish in their pond. All the development activities in the area where the land to be acquired by the Sahara have come to a stand still condition.

Conclusion

Apropos of the foregoing discussions it may be said that tourism industry is no doubt an important thrust to uplift the economy of an area. For the necessity of tourism, certain infrastructures have been developed and a change in the landscape has become obvious, introducing a new cultural landscape. But as Bakkhali is located in the Biosphere Reserve area, developers should be cautions to maintain the purity and serenity of the biosphere. Sea beach should be free from pollution made by the deposition of polythene bags and other waste materials, because, this pollution may cause disaster of living beings of the sea. A large number of people of this area are depending on fishing in the sea, they will loss their economy in the long run. Local authorities make known to the picnic parties and tourists about the detrimental reaction of throwing polythene bags and other waste materials on sea beach through continuous announcement near the sea beach. Authorities may inspire the local people to keep a watch over this matter.

Illegal collection of forest wood to supply hoteliers is also detrimental for the biosphere. A large section of population of this area is depending on the collection of shrimps (*meen*) as causing destruction of thousands of species of the sea. As a large portion of people are depending on shrimp collection and shrimps are necessary to develop the fish culture. It may not be easy to stop. The shrimp collectors should not be allowed to throw the other marine species on the sea beach. Rather, they should have those species in the sea water after collection of shrimp. To do this they may use some buckets with water to save the biodiversity.

The local authority should think about the sewerage system of the area. Waste materials should be burnt properly by the lodge and hotel owners instead of throwing them in the *nayanjuli* (roadside ditch) or *nullah*.

Table–10 : Occupation Pattern of Lakshmipur

Year	Main workers			Primary Sector			Secondary and Tertiary Sectors		
	Total	Male	Female	Total	Male	Female	Total	Male	Female
1981	138	129	9	106	103	3	32	26	6
	*21.66	20.25	1.41	**76.81	74.64	2.17	23.19	18.84	4.35
2001	242	178	64	61	47	14	181	131	50
	*25.80	18.98	6.82	**25.21	19.42	5.79	74.79	54.13	20.66

N.B. * Percentage to the total population
** Percentage to the total main workers
Source: Census, 1981 and 2001

Table–11 : Occupation Pattern of East Amrabati

Year	Main workers			Primary Sector			Secondary and Tertiary Sectors		
	Total	Male	Female	Total	Male	Female	Total	Male	Female
1981	844	831	13	472	469	3	372	362	10
	*25.22	24.83	0.39	**55.92	55.57	0.35	44.07	42.89	1.18
2001	1382	1183	199	264	192	72	1118	991	127
	*24.83	21.26	3.57	**19.10	13.82	5.21	80.90	71.71	9.19

N.B. * Percentage to the total population
** Percentage to the total main workers
Source : Census, 1981 and 2001

Selling of soil of agricultural land for high price to the developer of tourism industry is also very harmful to the nature, because, the agricultural land is becoming unworthy for cultivation. As a result the village will loose its capability to supply food to its people. They may use the soil of ponds/fallow land/canals (those have been choked due to natural deposition of silt/sand.

On the other hand, deterioration in social sphere due to addiction to alcohol, hemp, and gambling and sex business in the area cannot be ignored. These will destruct the economy of the poor village people. So, the local authority should handle those strictly by involving local people to eliminate those practices.

Although it was observed that some people are becoming aware to plant more trees in their surrounding areas to protect their land from salinity and encroachment of the sea. Thus, they are protecting their agricultural land; their houses as well as they are supplementing their needs of fuel. So, the local authority should encourage the activities of those people to the villagers at a large scale and inspire them to create more plants in their locality. Thus, green coverage of the area will be increased adding to the natural beauty. Simultaneously, forest will be saved from deterioration. It may be said that awareness of local people is very much necessary to save their natural as well as social sphere; otherwise, the notion of creation of Biosphere Reserve will be derailed.

References

Chawla, Romila, 2003, *Ecotourism and Development*. Sonali Publication, New Delhi.

De, Barun, 1994, *West Bengal District Gazetteers, 24-Parganas* Government of West Bengal Kolkata.

De, Rabindranath, 1991, *Sundarban* Pashchimbanga Rajya Pustak Parshad, Kolkata.

Debnath, H.S., 2002, *Sundarban Biosphere Reserve*. Botanical Survey of India, Kolkata.

Raha, A.K., 2004, *Indian Sundarban-An overview*. Director, Sundarban Biosphere Reserve, Forest Wildlife Wing, Forest Department, Kolkata.

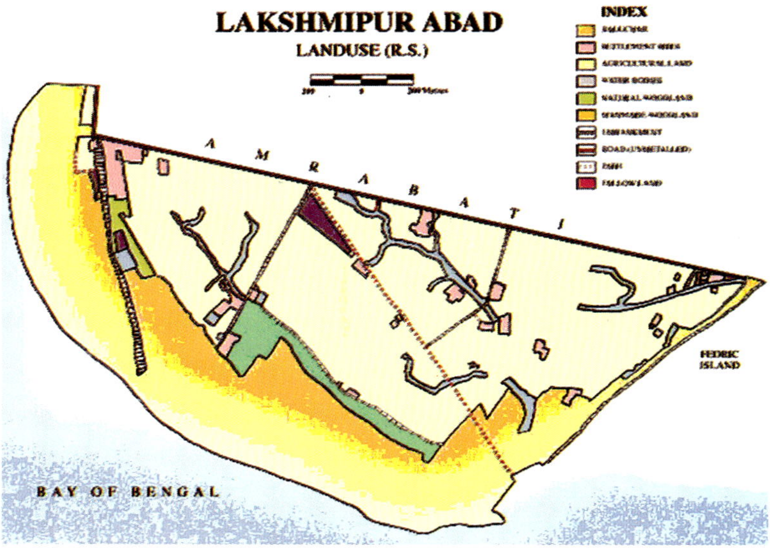
LAKSHMIPUR ABAD
LANDUSE (R.S.)
INDEX
A M R A B A T I
FEDRIC ISLAND
BAY OF BENGAL

Map 1

LAKSHMIPUR ABAD
PRESENT LANDUSE
INDEX
BALUCHAR
SETTLEMENT SITES
AGRICULTURAL LAND
WATER BODIES
NATURAL WOODLAND
MANMADE WOODLAND
EMBANKMENT
COMMERCIAL & INDUSTRIAL AREA
PROSPECTIVE COMMERCIAL AREA
ROAD (METALLED)
PATH
FALLOW LAND
A M R A B A T I
FEDRIC ISLAND
BAY OF BENGAL

Map 2

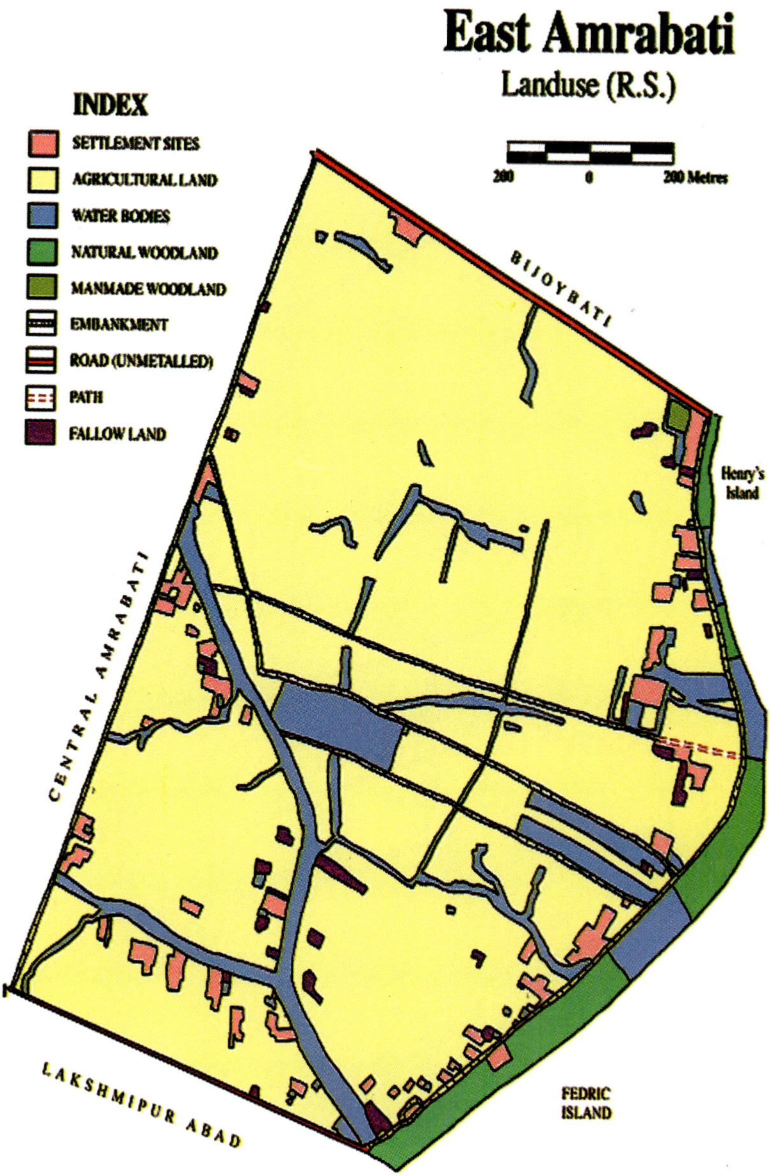
East Amrabati
Landuse (R.S.)
INDEX
SETTLEMENT SITES
AGRICULTURAL LAND
WATER BODIES
NATURAL WOODLAND
MANMADE WOODLAND
EMBANKMENT
ROAD (UNMETALLED)
PATH
FALLOW LAND
200 0 200 Metres
BIJOYBATI
Henry's Island
CENTRAL AMRABATI
LAKSHMIPUR ABAD
FEDRIC ISLAND

Map 3

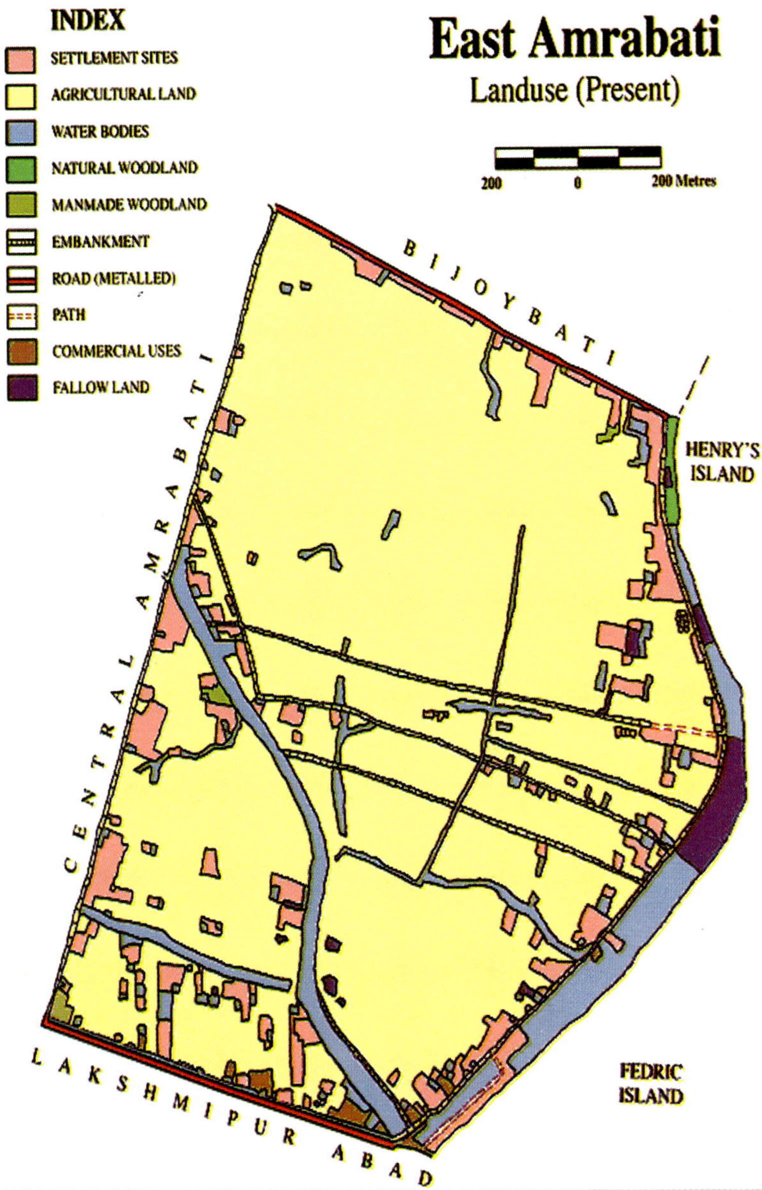
INDEX
SETTLEMENT SITES
AGRICULTURAL LAND
WATER BODIES
NATURAL WOODLAND
MANMADE WOODLAND
EMBANKMENT
ROAD (METALLED)
PATH
COMMERCIAL USES
FALLOW LAND
East Amrabati
Landuse (Present)
200 0 200 Metres
BIJOYBATI
HENRY'S ISLAND
CENTRAL AMRABATI
LAKSHMIPUR ABAD
FEDRIC ISLAND

Map 4

3

Structure, Cognition and Symbolism of Forest in the Folk Tradition of Sundarban

Kakali Chakrabarty

The folk tradition of Sundarban is unique in its own way .It portrays a large spectrum of cultural traits related to the unique environmental niche. Land, forest, water, wild animals make the immediate natural environment.

For any traditional society living in an isolated and extreme natural environment, the immediate environment shapes the economy and that in turn shapes its cultural tradition. In such a situation habitat, economy and the people enter into unique relationship, which consequently imposes and superimposes related ideas, practices, beliefs and rituals in the folk tradition. Tradition implies continuity and continuity does not mean stagnation. Rituals and practices change, old myths are reinterpreted in new perspective; and new myths are created and fitted into the existing framework of the old text.

With this initial understanding the present study intends to probe into the unique relationship between the habitat, economy and society in Sundarban the place having a unique physical environment of the world's greatest mangrove forest and abode of the fearsome man-eater Bengal tiger, identified as one of the world's most important Biosphere Reserves. Standard anthropological techniques of case history, interview, and observation with purposive and suggestive sampling have been adopted for collection of data. Data have been collected mainly from the forest goers. Some key persons with varied educational and occupational status were contacted for the basic information. Key persons included school teachers, panchayat members, health workers, local elites, and so on.

The Area of Study: Land and People

The unique forest tract of mangrove vegetation stretching along the sea face of the Bay of Bengal in the southern most part of the Gangetic delta is called Sundarban. It stretches across the international border of India and Bangladesh. During the British period Sundarban was extended from the estuary of the river Hooghli on the west of the 24 Parganas to the great river Meghna on the east of Bakarganj district (Hunter, 1876, rep. 1998:1). After partition of India in 1947, about two-third of Sundarban has gone out of India (the districts of Khulna, Jessore, Bakharganj, at present are included under the territory of Bangladesh). The Indian Territory of Sundarban stretches between the River Hooghli on the west and the international border with Bangladesh on the east, much of which runs along the river channel of Raimangal. It comes under the district of South 24 Parganas. The administrative boundary of Nadia district limits its northern extent and Bay of Bengal forms the natural boundary on its southern limit.

The name Sundarban, might have been derived from the *Sundri* tree (Heritiera litoralis), the most precious of timber trees available in Sundarbans. The tree is so named because of its reddish colour like *sindur i.e.,* vermilion. Though there are other propositions regarding the origin of the very name, but the one mentioned above is the most popular version. It has been reported from various sources that presently the number of *Sundri* trees, in the Indian territory of Sundarban, have declined sharply but they were undoubtedly plentiful earlier. Various written records are available in favour of this fact. To mention one is that of W.W. Hunter's Statistical Account of Sundarban (1876, reprint 1998) where he mentioned about the availability of stumps of *Sundri* tree at Sealdah and Fort Willium at Calcutta at various levels down to a depth of thirty feet (p. 7). A large number of *Sundri* trees were found in Canning, in Hooghly, and so on. All these certainly prove the abundance of *Sundri* trees in West Bengal. Both natural and man-made factors are responsible for its decline. Though typically mangrove vegetation, the *Sundri* tree requires sweet water for its growth. Earlier the sweet water of river *Piyali* was carried through the river *Matla* to this forest tract. Because of the gradual drying up of the river *Piyal* in course of time, the sweet water flow through *Matla* has been stopped, resulting into a sharp decline of *Sundri* trees in the Indian territory of Sundarban. Ruthless felling of *Sundri* trees since early days is another important reason for the degeneration of this species. Hunter (Ibid, 28) noted that, "a great part of the wood thus brought up from Sundarbans is *Sundri*." He found it necessary to close certain allotment, in order to allow the timber to come to maturity and to regulate a future supply (Ibid: 31). The other kinds of timber found in Sundarban are *Amur, Bain, Balai, Bhara, Dimal, Garan, Geona, Hental, Pasur, Kankra, Keora, Khalse, Kirpa, Sondal, Paras*, and so on.

The present study had been carried out in the Gosaba Block. The administrative limit of this block includes three islands namely Gosaba, Rangabelia and Satjelia with their headquarter at Gosaba. The name 'Gosaba' had been derived from the name of the river Guasaba or Gosaba, one of the main river courses that flow through this terrain. The present history of human settlement in this part of Sundarban is as old as of British period. The initiative to populate this land was taken up by the British Administrators with a definite idea to earn revenue for the government. Thus, the population of this area exlusively comprises the immigrants from different areas. In the earlier days of reclamation, these three islands used to constitute the territorial limit of Sir Daniel Hamilton's estate. Sir Hamilton, an Irish in origin, acquired three islands from the British government and established his own estate. Till date people remember him with great respect and pride. The most important contribution of Sir Hamilton was the introduction of the co-operative system and *Dharma Gola* or the public granary, one of which has been revived in recent years in Rangabelia. The new estate attracted people mainly from the adjacent areas of Khulna and Jessor. According to local sources (school teachers, panchayat members) a majority (roughly about eighty percent) in Satjelia Gram Panchayat is from the Khulna district. The present population of this area includes a number of castes, tribes and religious groups.

The reclamation work was mainly done by the tribal people brought from the Chhotanagpur Plateau areas and also from western districts of West Bangal. The tribal population comprises Munda, Oraon, Santal, Bedia. The tribals are locally called Adibasi or Sardar in general. It is commonly believed by the tribal people that they used to be the gang leaders (*Sardar*) of the labourers during the early days of reclamation and hence, in due course became known as *sardar*. It is to note that though their neighbouring caste people consider them under the single category of Adibasi, the tribals themselves maintain their individual ethnic boundary as Munda, Oraon, Santal, Bedia, and so on. However, it is observed that their original language is fading out. In Sardarpara, a small hamlet of Bedia community (having seven families) in Mitrabari village under Satjelia gram (village) panchayat, only two aged persons are found conversant in their original tribal language, which they call Bedia language. Now they are well conversant in the local Bengali language and among themselves they speak in a mixed language of Bengali and Bedia. Sri Dipchand Munda (Sardar) of Dayapur village submitted the same observation.

Besides the tribal people, the early settlers included the cultivators from Khulna and Jessor districts who were mainly from Paundra Kshatriya and Namasudra caste. The Paundra Kshatriya and Namasudra were the major landholding castes of this region. According to local panchayat sources Paundra Kshatriya are numerically dominant in Gosaba Block and the Namasudra are mostly concentrated in North 24 Parganas. According to Sujit Kanti Sur (1982:77-78) the Paundra Kshatriya might have hailed from the old state of Pundrabardhana of Barendrabhumi (old Rajsahi area). He opined that in some remote past the Paundras were compelled to leave their original homeland and migrated to Bengal. According to census record of 1951 the Pods of Bengal are all along included in the list of "depressed" or "scheduled" caste. They are considered to be remnants of an aboriginal tribe of the Ganges delta (p. 33). According to Risley (176-177) Pod are a fishing, cultivating, and land holding caste of lower Bengal. They are found in large numbers in 24 Parganas. The Paundra Kshatriya were locally called pod. Hunter (1876 rep. 1998: 36) also recorded them under the name pod. But the people themselves find the name pod as corrupted version of their original name of Paundra Kshatriya. The census of India 1951 also noted, "Pods now call themselves Padmarajas or Paundra Kshatriyas". The Paundra Kshatriyas in Gosaba Block are largely involved in fishing and forest based economy. Hunter in his statistical account (1998:36) also recorded them as cultivators, fishermen and occasional wood-cutters. But for fishing inside the forest area the Paundra Kshatriyas might have been a comparatively late entry than the Muslims of the area.

The Namasudras were recorded as Chandal in early literatures. The old literatures of Bengal like Dharmamangala, describes 'Bangal Desh' was inhabited by the 'Chandal'. They were cultivators. Hunter (Ibid) also recorded them as Chandal and described them as cultivators and occasional woodcutters. Sujit Kanti Sur (1982:78) described the Namasudra or the Chandal as the major population of southern Bengal who were later pushed towards eastern side of South Bengal by the Paundra. In the ancient literatures of Bengal (Charyapada No. 39) the Chandal had been referred to as the inhabitants of southern Bengal. According to Risley (I: 183) Namasudra are a non-Aryan caste of Bengal engaged for most part in boating and cultivation. Their number is largest in Decca division, Tippera, Jessor and Khulna but they are found in large number in 24 parganas, Nadia and Midnapur also. They are the second largest caste in Bengal.

The fishing communities of this area include the Rajbanshi and the Malo. Both the castes claim themselves as traditional fisherfolk or *jele*. The Rajbanshi and the Malo came to this area much later than the Paundra Kshatriyas and Namasudra. Rupa Mandal of Rajbanshipara of Gosaba traces their migration in the fifties of nineteenth. Tarak Mandal, aged 56 years, a resident of Hatkhola of Satjelia, who is a third generation descendant of one of the first group of settlers stated that the *jele* people of Emilibari village came only about 30 years back. The *jele* people of Emilibari also supported his observation. These later comers never tried to acquire land and continued with their traditional occupation of fishing. The Rajbanshi and the Malo, although not very prominent numerically but are important fishing communities of the area. They maintain a lot of their old traditions of folk songs from their original homeland of the then East Bengal. Bagdi is another forest based fishing community of the area, who claim them selves as jat-jele *i.e.,* the fisher-folk by tradition. According to one view, Bagdis are the earliest inhabitant of the area and the name might have been derived from the word Baghratati (from baghratati to bagri to bagdi). Baghratati has been accepted as the old name of Sundarban. Hunter described them as cultivators and fishermen and occasional woodcutters, numerous in numbers of whom, most are poor. The other castes living in this area include Jogi, Kaibarta, Napit, Kapali, Tantubai and so on. Jogis are a weaver caste by tradition and presently engaged in agriculture, trading and business. Apart from the immigrants from Jessor-Khulna area, people from Medinipur district of West Bengal also came in large numbers during the early days of settlement. They are mainly attached to land as informed by the panchayat sources of Satjelia G.P.

Besides these various castes of Hindu fold Muslims are also numerically dominant in Sundarbans. Muslims mainly live on forest based economy. According to the local people it was the Muslims only who used to live on forest from the very beginning of the settlement. They remained as woodcutters and honey collectors.

The Muslims of Sundarban called themselves as *sune musalman*. Hunter (1998: 36) classified them as Sunni sect of Islam. According to Sujit Kanti Sur of Gosaba , a veteran researcher on Sundarban. The word *sune* does not necessarily mean Sunni. He attached much importance on their pronunciation and infers that they were never formally converted to Islam at any point of history. They remained faithful to the *pirs* and *fakirs* who always proved helpful for these poor people and out of their sense of obligation they started professing the religion, which these Islamic preachers propagated. In due course they started to identify themselves as *sune musalman i.e.,* Muslim by listening or hearing to *Pir* or *Gazi*. *Sune* in Bengali is hearing to. The proposition is found acceptable as in none of the myths there is any mention of the formal conversion of the comon people rather it is described that the poor people spontaneously embraced Islam while the kings were formally converted to Islam after being defeated by the *Gazis (Gazi, Kalu, Champabati)*.

Apart from Hindu and Muslim there are Christian population who mainly engaged in occupations other than the forest based one. Of course exceptions are there. Hunter did not mention much about the Christians who were very few in numbers during those days.

Early accounts spoke of the Malangi, the salt manufacturers of Sundarban. According to Hunter (rep. 1998: 37) "the principal immigrants to the 24 Parganas Sundarban are Uriyas from Cuttack and other parts of Orissa and Bunas from Chotanagpur or Western Bengal. They originally came as Malangi or salt manufacturers and as woodcutters". Since the salt manufacturing centers of the 24 parganas closed down permanently due to a ban imposed by the British government on all kinds of salt manufacturing in 1896, the Malangis lost their job. Hunter (Ibid) stated that many of them have settled down as cultivators in various parts of Sundarban.

For the development of permanent human habitation in the low-lying tract of Sundarban, the first and foremost task was to raise high embankment (*gher*) around the reclaimed islands along with the river course to ensure the protection from the saline water of the rivers. This may be mentioned here that the level of the land in this area is much lower than the level of water during the high tides. One interesting feature of this part of Sundarban is the salinity of its river water. During the high tides sea water enters into the rivers and estuaries, leaving the water saline.

Land, water and forest form the basic economic resources. Land promotes agriculture. Forest provides honey, wood and wax. Water resources promote fishing and catching of other aquatic animals. Because of the salinity of river water agriculture depends on rainwater only, thus producing single crop in a year. With the limitations of the growth of agriculture in full swing, forest remained the major economic resource since the very beginning of the human habitation in and around the area.

Historicity of the land: British and pre-british period

Sundarban, demarcated had been as a distinct territorial unit by the British Administrators in order to streamline their administrative jurisdiction over the land. As early as in 1807, the British Rulers felt the need of statistical survey of this country for their own administrative interest. In 1867, the Governor General in Council directed a statistical account to be prepared for each of the twelve great provinces of India of which Bengal was one. The accounts were prepared district wise and Sundarban was treated separately from 24 Parganas. However, under the British Rule, Sundarban never formed a separate district with its independent administrative set up, although there was a local commissioner to look after the the revenue matter. In 1819, the jurisdiction of the Commissioner was extended and he was authorized, with the sanction of the government, to grant leases of forest land within the tract belonging to the Government (Hunter, 1998:1-2).

The early accounts of the British administrators (Westland; 1998:2) described the reclaimed lands of Sundarban in the Jessor district without having any settled villages. The cultivators used to live far apart from their rice field. It was under the British Rule that planned reclamation work started afresh with the initiative of Mr. Henkel, the first English Judge and Magistrate of Jessor as early as in 1784.

Although the present history of Sundarban begins with the British period, there was even earlier history of human settlement in this land, which is evidenced by oral traditions and by sporadic archaeological findings from various places of 24 Parganas. In fact, the archaeological findings from various places of Sundarban prove human habitation in this area even before the Muslim period. Jalil (2000: 44, 48) stated that before the Muslim rule, this tract was inhabited with the Buddhist people and some aboriginal people like Paundra Kshatriyas or the Pods and Namasudras or the Chandals. From the copper plate inscription found at I'dilpur in Bakarganj, many historians are of the opinion that, this part of the land was inhabited with a tribe called Chandabhanda, who were salt manufacturers. The plate recorded the grant of three villages by Madhava Sena, the king of Bengal, to a Brahman who, with landlord rights, received the power of punishing the Chandabhandas or Shandabhandas, a race that lived in the forest (Hunter, rep.1998: 103). It is also evident from various archaeological findings that Sundarban was a center for manufacturing salt since very early age and local tribal people were involved in it. H. Beveridge (c.f. Hunter; 1998: 117) wrote that 'it seems evident that in old times salt was reckoned as the most valuable production of this part of the country.' Eye witnesses reported that deep into the forest of Haldi and Chamta area certain common domesticated plants like *Jeol, Banyan, Jam* (Black Berry) are present, which certainly prove the earlier human habitation. These trees are not natural to the mangrove vegetation. Jalil (2000:49) reported similar observation in some other areas of Sundarban forest. All these prove that Sundarban had been populated and depopulated over and again. It is also a fact that not all the parts of this land had been populated at a time. Natural calamities, geophysical factors, invasion of the Magh, Arakanese and the Portuguese pirates are believed to be some probable reasons for depopulation of the country. The changing course of the main flow of the river Ganga is another important natural reason.

The earliest attempt to promote human settlement in Sundarbans has been traced back to the first half of fourteenth century by one namely Khanja Ali or Khan Jahan Ali, Muslim chief who died in 1459 A.D. (Zill Hajj 863) at Bagherhat. In Khulna Gazetteer (1914) O'Mally wrote "the earliest traditions of the district are connected not with any ancient Buddhist or Hindu Kingdom but with a Mohammedan called Khan Jahan Ali or more generally Khanja Ali. Local legend relates that he came here over four centuries ago to reclaim and cultivate the Sundarbans which were then waste and covered with forest." Historically much is not known about him. After a thorough academic exercise Jalil (2000: 246) concluded that Khan Jahan was of Turk origin and he came to Sundarban from Delhi with huge assets, which he invested for the welfare of the poor people of the region. His original name was Ulugh Khan Jahan and was titled as Khan-e-Azam. Ali might have been a later added title. He was elevated to a status of a Pir (Muslim saint) by the people. According to Wasteland (c.f. Jalil, 2000: 46) between the period of 14th and 16th centuries many Muslim Pir and Missionaries came to Jessor to propagate Islam. Beveridge (c.f. Hunter, 1998:122) concluded 'there is nothing in this contradictory to the fact that Jessor formerly belonged to Khanja Ali [Khan Jahan]... It is true it had been cultivated in the previous century by Khanja Ali, but the experiment had proved a failure, and the land had in the time of his successor (?) Chand Khan replaced into jungle.'

Besides the story of Khan Jahan, the oral tradition of Sundarban is also rich with the stories of *Gazi, Pir* and *Fakir (Faqir)* who came to Sundarban at different ages and influenced the local people greatly with Islamic preachings. They won over the heart of the poor with their various welfare services. James Wise (c.f. Hunter, 1998:129) observed "when the Muhammedan armies poured into Bengal, it is hard to believe that they were not welcomed by the hewers of wood and drawers of water, and that many a despairing Chandal and Kaibartta joyfully embraced a religion that proclaimed the equality of all men, and which was the religion of the race, keeping in subjection of their former oppressors." *Gazis* were actually the soldiers who under many compulsions came to take refuge among these tribes in these inaccessible areas and gradually became popular with their welfare activities. In due course, they earned regards as *pir* and became the legendary figures. Mention may be made of Zinda Gazi, Pir Mobarak Gazi, Barkhan Gazi, Pir Gorachand, and so on. They are remembered through the oral tradition of myths and tales. These myths and tales act as pivotal in the development of the unique cultural identity of Sundarban. The analysis of these oral traditions is not only important in understanding the folk tradition of Sundarbans; it may also unveil the unrecorded history of the land.

Identity of the land through ages: *Bada, Badabon* and *Bhatidesh*

The name Sundarban, for this tract of land became popular mainly during the British period. Earlier this land of lower Bengal was known as *'Bhati'* or *'Bhatidesh'*. This empirical data was supported by various accounts of various scholars. Hunter (re 1998: 103) also observed, "the extension of the name of Sundarban to the whole coast is evidently modern. The Muhammedan historians do not use the term, but give (as Abul Fazl in the *Akbarnamah*) the coast strip from Hijli to the Meghna the name of *Bhati*, which signifies 'low lands overflowed by the tides". He further observed that even during his days the term *Bhati* was still in usage in Jessor and Bakhargunj district for Sundarbans and the inhabitants of the area were referred to as *Bhati-lok*. Jalil (2000:233) also noted that Abul Fazl in *Ain-e-Akbari* referred to the eastern part of Bengal as *Bhati*. During the reign of Akbar, the Mughal Emperor, this country had been referred to as *Baro-Bhatir Bangla i.e.,* Bengal of twelve *bhati* , after the famous *Baro-Bhuinya* of Bengal (Jalil, 2000:13). In the oral tradition of *Bonbibi* the land has been referred to as *Atharobhatir desh i.e.,* country of eighteen *Bhati*.

According to Jalil (2000) the region where low tide regularly occurs in rivers is called *Bhatidesh*. The people of the area also viewed the same. According to one explanation (in empirical situation) the area has been referred to as *Atharobhati* because of the fact that if one travels from the first bhati to the last one by waterways it takes eighteen ebb tides to cross. In earlier days when waterways was the only means of communication in Sundarbans the boatmen used to refer the distance between the places in terms of number of Bhati or ebb tide they had to cross like *char bhatir rasta* (a distance of four ebb tide), *chhay bhatir rasta* (a distance of six ebb tide), and so on. The present author logically derived a different opinion. According to the present proposition the word *Bhati* here denotes a distinct territory separated from the other territories by physical or arbitrary boundary, otherwise called as an estate language or *taluk* in local. Thus, the land was called *Baro bhatir Bangla* after the twelve estates of twelve landlords (*Baro Bhuinya*) of the then Bengal or *Atharo-Bhat* in *desh* after the eighteen estates under the mythical rulers like *Dakshin Rai, Barkhan Gazi, Bhangar Shah* and so on. The myth also tells about one particular *bhati* namely Kendokhali, the only *bhati* left with *Dakshin Rai*. In reality, Kendokhali is the farthest island in the south on the sea. In colloquial language also *Bhati* is a common word for a well-defined territory for specific purpose, like *chun-bhati* (lime-kiln), *int–bhati* (brick-kiln), *nun-bhati* (salt–kiln) etc. It may be mentioned here that according to Sri Sujit Kanti Sur, a veteran researcher on Sundarban, *Bhatidesh* stretches between the eastern bank of the river *Matla* and the western side of the river *Kabadak* (now in Bangladesh).

Another identity of this land is *Bada or Badabon*. *Bada* is the local word for low-lying marshy land in Jessore-Khulna dialect, and *Bon* is the word for forest. Till today the local people prefer to call their land as *Bada* or *badabon*. The people from the northeren part of South 24 Parganas refer to the southern part of the district as *Abada* or *Abad*. To their vocabulary the reclaimed low land transformed into arable one is called *Abad*. According to the villagers, the estate of Sir Daniel Hamilton, consisting of three islands namely Gosaba, Satjelia and Rangabelia, was popularly known as '*Hamiltaner Abad*' *i.e., Abad* of Hamilton. These two terms, *Bada* and *Abad*, are most common in usage till today by which the local people of Sundarban demarcate themselves from the 'other' people.

Sundarban: A cultural area

As has already been stated in the previous section, the oral tradition of Sundarban is rich with the myths and stories of *Gazi* (Muslim warriors), *Pir* (Muslim saints) and *Faqirs* (Islamic mendicants). These oral traditions give a distinct and common identity to the whole tract of Sundarban across the countries. The stories of Zinda Gazi, Pir Mobarak Gazi, Ali Madap, Bonbibi, Shah Jangali, Barkhan Gazi, Dakshin Rai carry such unique identity of the common tradition. Considering this distinct and common cultural tradition the present study proposes Sundarban as a cultural area, having folk groups adaptive to its typical cultural environment. The empirical findings also reveal a feeling of commonness among the people over this vast tract of land named Sundarban. The tradition they share includes the language, beliefs, customs, rituals, folk songs and tales that have been developed uniquely in a peculiar ecological niche consisting of forest, rivers, estuaries, island and wild beasts precisely the man-eater Bengal tiger. Besides these stories of *Gazi, Pir* and *Faqirs,* the typical tunes of *Manasa Bhasan, Sitala Jagoran, Dukhe pala,* carry the unique cultural flavour of this folk tradition.

It is important to note that not all the characters discussed above earn equal importance in all the places at a time. Each character has a finer territorial limit of prominence. Thus, the myth of Bonbibi occupies a central place in the eastern part of the river Matla, that of Dakshin Rai predominates over the western part of the river, Pir Mobarak Gazi remains the central character in the areas further north, the story of Zinda Gazi is pivotal in the area in and around Maimansingha (now in Bangladesh). Based on the prominence of a particular cult-figure the cultural area of Sundarban can be divided into a number of sub-cultural zones. The present Indian territory of Sundarban can be divided into two sub-cultural zones on the basis of the predominance of *Bonbibi* and *Dakshin Rai* cult. The eastern part of the river Matla may be classified as the Bonbibi cult zone while the western bank of the Matla comes under *Dakshin Rai* cult zone. According to local elites the eastern part of the Matla river is least invaded by the mainstream people and thus retained their cultural tradition in their original form. The economic dependency on forest is much more in this zone.

Structure and Cognition of forest in the folk tradition

The people of Sundarban refer the forest as *mal*, a colloquial usage for the term '*mahal*' in Jessor–Khulna region (now in Bangladesh). The word *mahal* in Bengali means an estate, a portion of a total estate or the home or abode of someone. The term was in common usage, specially, during the feudal system. The forest tract was usually referred to as *jungle mahal*, as a part of the larger estate. Under the Muslim Rule in India revenue divisions in Bengal were called '*mahal*'. Under the Mughol Rule (during the time of Akbar) Bengal was divided into 19 *sarkars*, containing 689 *mahals*. Under the control of Shah Suja, Bengal was divided into 34 *sarkars* consisting of 1350 *mahals* or *parganas*. Later small *mahals* were clubbed into *chakla*, and *parganas* (Hunter rep. 1998: 77-78).

The common people of Sundarban, however, do not have much idea about the origin of the word; their idea of *mal* is fixed with the forest only. In the cognition of the people the forest is associated with supernatural and magical beliefs, practices and prejudices. Myths, particularly of *Bonbibi, Dakshin Rai, Gazi* and *pir* play an important role in their understanding of forest. Forest is understood not in simple terms of man-nature relations but in a complex set of interaction between man, nature and supernatural power. The myth of *Bonbibi* and *Dakshin Rai* is in the central place in this man–forest relationship. Here myth is an encapsulated form of their past. Their interaction and relation with the forest is explained in the light of the myth only. Thus, a tiger is not simply a beast but he is *Dakshin Rai*, the defeated king of *atharobhati*, a supernatural power, who needs to be appeased; forest is not simply a piece of land with trees but is the estate of *Bonbibi,* the divinity and therefore should not be polluted by spitting or defecating; a boat is not simply a wooden structure but is the abode of *kasthadevi* who alarms them prior to any awaiting danger; a good earning of honey, a good catch of fish, safe return from the forest–all depend on the mercy of *Bonbibi*. The cause of failure is the responsibility of the person concerned but success always added to the credential of the deity.

Myths make the ground of the belief system, beliefs are manifested through the rituals, ritual get expression through symbols, which in turn are regulated by the myth.

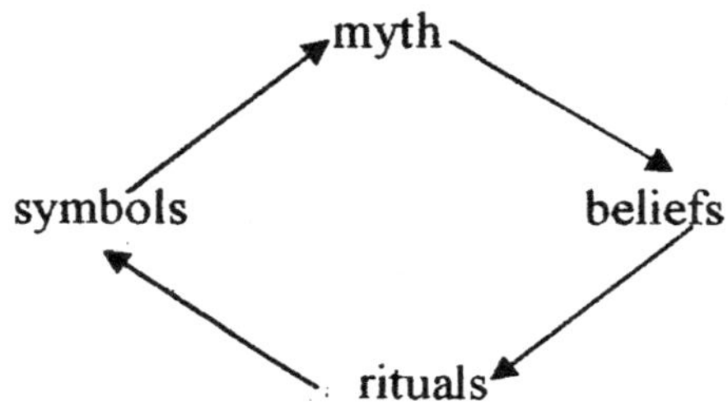

Fig. 1.

The relationship between *Bonbibi* and *Dakshin Rai* cannot be conceived in simple terms of winner and defeated, but with some humanely sentiments where the myth leaves ample scope for the storyteller to incorporate his own feelings and interpretations. Thus, though the myth says that *Dakshin Rai* was allotted one *bhati* namely *Kendokhali* only but people interpret that along with *Kendokhali* he was given the overall responsibility of protecting honey also, hence he is to be appeased along with *Bonbibi*. Interpretations vary from person to person and the process of creation and recreation of new myth is on move.

The story of Aswini Hajra, a fisherman by occupation, may be cited as an example. After receiving a message from *Bonbibi* in dream, Aswini ventured in a particular spot for fishing where he got plenty of Hilsa fish .The supply of fish was so plentiful that not only Aswini but also other fishermen made exorbitant profit. Aswini was so grateful to the deity that he installed an icon of *Bonbibi* on that particular place in the forest and the very spot became famous after his name as Hajrakhali. The tale is very popular among the fishermen and they take liberty to incorporate new characters and incidents.

The cognition of supernatural works in two fields, one is the religious field associated with organized pattern of rituals and the other is the magical aspect of it performed at the individual level through magic verses. In folk tradition they are in a continuum rather than compartmentalized. Thus, the belief of *Bonbibi* carried on to the midst of the forest through magical rites. The organized and collective form of worship is performed on the riverbank *i.e.*, on the borderline of the forest. As soon as one enters into the forest the world of magic begins. A *baule* or *gunin* is the performer of the magical rites and he is believed to have precise quality of invoking divine power through *mantra* (magic spells). These mantras have enormous importance in the life of the forest goers. Mantra cannot be uttered without any cause. It looses power if it is spelt out publicly. Mantra can only be transmitted from *guru* (teacher) to a *sisya* (student) on payment of *guru-dakshina* (honorarium). *Mantra* works only in the midst of the forest.

Mantras are for different purposes. An expert *baule* could smell the danger as soon as he touches the ground in the forest. There are two folk expressions opposite to one another. Suspected presence of tiger in the working allotment is called '*mal garam*' and just opposite situation is '*mal naram*' or '*mal thanda*', meaning the area is free from tiger and is safe. The forest goers compulsorily avoid uttering any word that is inauspicious to their knowledge. Thus, a tiger is referred to as '*bare miyan*' (the Pig person), '*raimoni*' (the dear king) and so on. In case a person falls prey to the tiger it is told that the person has fallen down into the forest (*jangale porechhe*).

Besides these religious and magical practices the forest goers also depend on their senses (on eyes and ears–in their terms), alertness, rationality and strength. They know from their age long association with forest that alertness and precise ability of observation are the prime qualities which a forest goer must develop within oneself. This knowledge and understanding of forest is to be transmitted from old to new, from experienced to novice. According to Phani Gayen, a *gunin* by himself, a good *gunin* is one who is brave and is able to make a decision promptly. A *gunin* is supposed to lead a team; he should be the first person to get down from the boat, be the last person to escape and should be at the tail end while leaving the forest. The responsibility of the safety of the team lies on him. According to Phani Biswas, a veteran forest goer, that unity and courage are most important in forest. He believes that it is the experience and knowledge of a *baule* on which the team rely and it is the unity and courage of the team on which the *baule* depends. Surhid Sardar of Dulki village informs that the new entrants in the team are taken care of by the experienced people. He believes that the mantra of a *gunin* fails only when there are someshort comings from his part like his attention is diverted to some other things, or he could not spell out the mantra in the right order or some other reasons. If the tiger's legs are not directly on the ground the mantra may not work. With all loyalty to *Bonbibi* he also feels alertness and bravery are most important qualities for forest goers. Thus, myths, supernatural beliefs, practices coupled with man's experiences, knowledge and wisdom form the very basis of structure and cognition of forest in the folk tradition of Sundarban.

Forest economy, associated beliefs, practices and traditional wisdom

Forest as economic resource act in two ways, one is direct way of interaction and the other is indirect way of interaction. For woodcutting and honey collection, people get direct exposure to the forest ground, but for fishing and crab collection people depend indirectly to the forest as they exploit the aquatic resources of the forest. It is important to mention here that nowadays the fishermen depend on the water resources of the forest territory as the narrow canals (*khal*) and the streamlets produce good yeild. On the contrary, the rivers in the open areas do not yield enough catch, as observed by the fishermen themselves. A number of factors are responsible for the decline of water animals in the rivers. One of the important factors is the growing occupation of catching of the tiger prawn larvae, locally called *Bagda meen*. Bagda meen are collected by *tana jal* (the type of net that is pulled through the water). While netting, along with the Bagda meen, other fish seeds are also trapped which go waste. Moreover, there is a growing rush into the field from all sections of the people including women and children. This has a tremendous negative impact on the yeild of fish and other water animals in the rivers. Again, based on the frequency of exposure, forest expedition may be classified into three categories, regular, seasonal and occasional. It is the fishermen who can be categorized as the regular forest goers. Earlier the woodcutters used to be the regular forest goers. But with the ban on woodcutting and the strict measures taken up by the forest department, the act of woodcutting has been stopped, leaving the woodcutters jobless. The withdrawal from woodcutting causes further rush to the aquatic resources of the forest area, causing an imbalance in man-resource ratio. The statement frequently came from the fisher-folk that forest area, nowadays, is more crowded than the villages. in reality it may not be so but the statement certainly gives a close view of the situation.

The seasonal forest goers include the honey collectors, and the occasional forest goers, are those, who take a chance to earn hard cash in need or time to time. The woodcutters used to be called as *bauli* or *bawali* and the honey collectors as *maule*. In the sixteenth century literature namely in *Chandimangal* by Kabikankan Mukundaram (1975) the woodcutters were referred to as *Berunia* or *Beruni*. In due course, the implication of the word *baule* has been changed to mean the *gunin* who possess the power of magic spells. *Baule* and *maule* are not a caste by itself but an occupational identity, the membership to which is flexible. A *maule* or *baule* may be from any caste or community.

One important aspect of forest-based economy is the group effort. The incongruous environment of the forest demands a group or teamwork and a strong fellow feeling among the teammates. Thus, a 'team' is most important in forest based economic organization.

To the fishing party the team is called *'sain'*. A *'sain'* is led by a *'saindar'* who must be an experienced person in the field. A '*sain* may be as small as of three members and as big as twelve members or more. Earlier a '*sain*' often consisted of two, or three or even more number of boats, each boat having three to five members. Nowadays, the size of the team has become smaller consisting of three to five members. Earlier the Malo and Rajbanshi who claim themselves *jat-jele i.e.,* the traditional fishermen did not opt for forest resource. But nowadays, a whole days labour hardly produces a sufficient earning and the fishermen find no option but to rush to the narrow canals inside the forest as it still gives them a good return.

Frequent exposure to the forest areas and involvement of more and more people into the forest-based economy resulted into keenness on smaller team. The members of the team often changes but it dose not affect the fellow feeling of a team. The teams do not bother for caste and creed of its members. Thus, a Muslim does not feel secluded on any matters even when the team mates are non-Muslim including Hindus. While in forest they cook together, share food, and share the same beliefs and practices. This fellow feeling developed a congenial relationship between the communities at the social level. Each team is compulsorily accompanied by a '*baule*' or '*gunin*', the worshiper of *Bonbibi* and the performer of magical rites in the forest for the welfare of the team. Nowadays the want for a professional *boule* is declining as often any of the teammates is found conversant in the *mantra* (magic spell) and he performs the magical rites for the team. This is often preferred from the monetary point of view. There is a definite norm of sharing the earning from an expedition. The catches are shared equally among the teammates including the *gunin* or the *baule*. The boat-owner and net-owner also have an equal share each. Thus, if a member of the team owns the boat, he would receive two shares, one for himself and one for the boat, likewise if a member posseses the net he would receive two shares, one for himself as a member of the team and one for the net. But if a member performs the magical rites he does not receive any extra share, only the team could save the share of the *baule*. Often the team members do not posses the net by themselves, and it is hired from *mahajan*. A *mahajan* is one who professionally lends out cash and kind to the fishermen. This way the *mahajan* takes entry into the economic organization. They lend out their net and other accessories to the fishermen on specific conditions on their catch. The *mahajan* also lend them cash with which they manage the expenditure of their family, of permits from the forest department and of ration for the team during their stay in the forest. This way the fishermen come in bondage to supply their catch to the *mahajan* on the rate fixed by the *mahajan*. These *mahajans* are the *khatidar*, the owner of the *khati*, the fish depot. It is the *khatidar* who actually control the fish market. *Mahajan* also receives shares for net and other accessories. Till the loan of the *mahajan* is repaid, which is somewhat impossible job for the fishermen, they cannot go to any other *mahajan* and this way they are trapped in a permanent kind of bondage.

The forest office issues the permit (pass) against a fixed rate of fees per individual in a team. The areas are well defined by the forest department as core and buffer zone. In local terms the core zone is called '*bandha*' (closed) and the buffer zone is '*khola*' (open). According to fishermen the '*khola*' zone does not yield sufficient catch due to over rush of people. Nowadays women are rushing into the occupation of fishing particularly in catching of *bagda meen*. They prefer to work in '*khola*' areas. Women normally do not stay back in the forest area at night. They go for single days venture. During the present study the author could witness three expeditions where the team comprised exclusively of women. Sometimes a family team includes its female members. The cases of Gita Mandal and Subhadra Sana of Jelepara of Emilibari village may be cited here. Gita is a widow. Her husband fell prey to a tiger. Gita was given shelter by her parents in her parental house. When her husband died she was pregnant by five months. She was having two more daughters. So after giving birth to the third child she found it wise to accompany her father for fishing. Her father also agreed upon her proposal, as it would not be the first case in the community. Since then Gita became a regular forest goer. The case of Subhadra Sana does not differ much. Subhadra goes to forest with her father and husband. Her husband is handicapped. In an expedition he got an injury in hand for which the thumb of his right hand had to be amputated. Subhadra accompanies him thereafter. Rabin Mandal of Satjelia informed that nowadays women do not dare to enter into the forest. He came across one such team very recently. In recent past one woman aged about 50 years, was attacked by the tiger while she was picking up the bagda meen. Other women could drive out the tiger but the poor woman died. Her body was brought back to the village. The incident was not reported to the local authorities because of the fact that the expedition was made without the knowledge of the forest department. Such incidents are not rare. The fishermen, who enter into the forest with legal permit from the forest department, get the benefit of insurance coverage on their life. If a permit holder dies of an accident in the forest his family gets a sum of rupees two thousand as compensation from the authority.

The forest goers often enter into the core zone for a good catch of crabs and fish and often are caught by the forest officers. During the present study one such case was reported. The involved party was charged with a fine of rupees one thousand in cash and their net and boat were ceased. The families involved had no other source of income and were living on meager diet. Under such circumstances these people have no other option but to go for an expedition without any legal permit.

There are three main items a fisherman looks for: fish, crabs and prawn larva. A fishing party enters into the forest during *joar i.e.,* flow tide as the narrow canals inside the forest are navigable only during *joar*. In rising tide with the flow of water, fishes swim into the canals. During *bhata i.e.,* the ebb tide water drains out leaving the fish into the net. During this ebb tide the holes of the crabs get exposed and it is the high time for crab hunting. The supply of *bagda meen* rises up during the flow tide. Thus, tide and ebb have enormous importance in the life of the forest goers. During the high tide women of every village are found catching *bagda meen* in the river near settlement. Children also participate in large numbers. Women in the settlements go for tiger prawn larva even in the midnight depending on the time of high tide. Debala, wife of Debaranjan Mandal of Emilibari village, informed that she goes for *bagda meen* even at 2.30 a.m. if it is the time for *joar*. She also said she goes together with other village women always, a fact reveals the bare reality of life. One should not overlook the fact that Sundarban-rivers have *kamoth* (small shark) and it attacks people. Accidents are not rare. The operation continues for about six hours *i.e.,* the whole duration of a rising tide till its pick and Debala earns about rupees 20 to 25 on an average. The working hour for fishing is controlled by the natural occurrence of tide. The high tide and ebb tide occur every six hours alternatively. Thus, there occur two high tides and two ebb tides in 24 hours. The fishermen come out of the forest during *joar* only and they prefer to anchor their boat on the open river. A full time fishing expedition continues for about ten days, from the eleventh day of a lunar fortnight to the sixth day of next fortnight. The fishermen usually fall prey to tiger at the time of picking up of fishes from the net or at the time of crab hunting. This extreme uncertainty of life demands a team; togetherness is the only strength for them to fight out every odds.

Honey collection (*mau chhanta*) is purely a seasonal activity as honey is available in particular season depending upon the blooming of the trees. The period between the month of *falgun* (Feb–Mar) and *Jaistha* (May-Jun) is the main season for honey collection. Two types of honey attract higher rate. They are of *khalsi* flower and of *garan* flower. According to expert *maule*, the honey of *khalsi* flower is lighter in shade (*sada i.e.*, white) and is called *phulpatti madhu*. This variety of honey is most popular. The honey of *garan* flower is dark red in colour and is sold in high price. Other kinds of honey includes *banihar madhu i.e.*, honey of Bain or Keora tree, *bogra madhu i.e.*, honey of Hental tree. *Bogra madhu* is scented. The honey of Keora tree is slight bitter in taste. Nowadays, forest department has introduced boxes for bee keeping in the villages. It produces good quality of honey. In the boxes the honey produced is of *sajne* flower. It is also getting popular in the villages. Wax is available along with honey.

Honey collection is also a group activity. One of the team members keeps watch on the movement of the bee. When the bees fly along a lower horizon it means it is carrying honey in its stomach. Sometimes bees fly just over the water level and people could get the message of flowering of trees in the forest. In the forest the *maule* minutely observe the movement of the bee to trace out the bee-hive. This knowledge they acquire through exposure to the forest and in association of the experienced people. Though honey collection is a seasonal activity but a major source of earning.

Mou chhanta i.e., collection of honey is taken up as a subsidiary source of income. Forest department issues license for honey collection. From the forest department masks have been provided to the honey collectors. All the forest goers compulsorily worship *Bonbibi*. The very first collection of honey is offered to the deity.

As soon as a team enters into the forest the very first activity is to worship *Bonbibi*. The *baule* is to get down first from the boat. The worship of *Bonbibi* is followed by the sacrifice of a fowl. Sacrifice is symbolical and the fowl has actually been freed into the forest in the name of the deity. Along with *Bonbibi, Dakshin Rai, Barkhan Gazi* and *Ali Madap* are also worshipped. The worship is followed by *dohai* (saying of name of the deities). With this one enters into the world of magic. *Dohai* is the last ritual performed jointly. After that the world of magic spells start. A hut is compulsorily made for *Bonbibi*. Either the icon or the earthen mounds are erected symbolizing the deities. Altogether five earthen mounds (*than*) are erected, one each for *Bonbibi, Dakshin Rai, Barkhan Gazi* , *Shah Jangali* and *Ali Madap*. Offerings (*hajat*) are made with *batasa*, (sweetmeat) jointly. In case any accident occurs, the party immediately leaves the forest and places a white flag to alert the other teams. The magical rites, performed in the forest to make the allotment free of tiger is called *mal kara* (meaning doing *mal*). It is reported that women are tabooed to perform the rites of *mal kara*. They are also prohibited to either magic verses (*mantra*). Hence they depend on the prayer only. The most essential part of the *mal kara* is to place three leaves (of any tree in the forest, preferably of *emo*) on the ground and keeping the right hand on the leaves the *baule* utter the magical verse. In doing so a *Baule* can feel whether the mal is *garam* or *thanda*. The forest-goers sincerely believe in the power of magic spells.

Besides the worship of *Bonbibi* the forest goers also worship *kastha devi*, the goddess of boat and *ganga devi*, the goddess of water. A boat is believed to be the abode of *kastha devi*, who alerts them from any awaiting danger. The boat makes a typical sound, which originates from the *golui*, the frontal end of the boat, believed to be the seat of the deity and passes on towards the back end. In their term it is *nouka daka*. A boat warns (*dake*) because of polluted water flow in the river or of certain awaiting danger in the forest. The first step is to sail the boat away from that very spot. If the sound does not subside then they change their destination and move to some other allotment of the forest. They believe that it is the *Kastha Devi* who alerts them. Uttara Biswas, wife of Pagal Biswas, a resident of Gosaba village no 3, worships a piece of wood, a broken part of *golui* of their old boat, as *Kastha Devi*. She narrated a long story. The boat was lost twice, once it was floated away with strong wind and on other occasion it was drowned into the river in a storm. On both the occasions the goddess appeared in dream and asked them to bring her back. She only showed the way. Both the times they discovered the boat at some convenient place to bring her back. Lastly they had to repair the boat anew and preserved this broken part. The part of the boat is placed beside the *tulsi mancha* (the holy place planted with a *tulsi* plant) and worshipped every day. Though in no other instances such regular worship is observed but the fishermen compulsorily worship the goddess on the boat before setting the boat on sail.

In recent years *Ganga puja*, the worship of goddess *Ganga*, is gaining importance in fishing community. The women, who are engaged in catching of tiger prawn larva, mainly perform this worship. Gradually it is taking a shape of community worship. The concept of *Ganga* is not new to the community as it is compulsorily performed on the boat. What is new is the introduction of an icon and transform it in a festive form. Traditionally no icon is used for the worship of *Ganga Devi*, the fishermen simply make an offering in the name of the deity and pour it into the river. Such rituals are performed exclusively by the men and on the boat before it sets on sailing (already discussed). Earlier, *Ganga Devi* was not worshipped at the village level also. Nowadays she is worshipped with all pomp and show in the village and the priest is compulsorily a Brahman (the priestly caste). The icon is typically of Hindu mythology, the deity sitting on *makar* (Capricornus, a mythological marine animal). A new myth is under process of creation around the goddess. The typical motifs of deity's appearance in dream, miraculous escape from the danger are coming up around the very concept of *Ganga Devi*. A folk tradition can never be static. It continuously creates and recreates myth. Now the icon of *Ganga Devi* is preserved in the *deora* (the place of worship), sometimes common for all worshipping deities.

Nowadays it is a remarkable change that worship of *Bonbibi* is not limited to the forest only. It extends far beyond the level of a village *deora* to the household level. The economic deity is slowly transforming into a household deity. It is observed that *Bonbibi* is worshipped even for the want of a child by the issueless couple. Example may be cited of Subala Mandal, the wife of Haripada Mandal of Pakhirala village. Haripada was a forest goer. He stopped going to forest by order (*hukum*) of Bonbibi. Afterwards Subala got a stance from the deity asking her to worship her everyday. She constructed a *deora* on her homestead land. She receives the message from *Bonbibi* under stance. People come to her *deora* asking favour of the deity on occasion of sterility, fever, ill health etc. It is believed strongly that all these problems can be solved with the favour of *Bonbibi*. A case study has been recorded in the present study that a young woman came to the deora to offer *mannat* (vow) after having an issue. The couple was childless even after eight years of conjugal life. Then they came to Subala with a request to pray the favour of the deity on behalf of them. They earned a strong faith on the deity through their own instance.

The participation of women in the rituals is an important change in the tradition. Earlier females were not allowed to participate in the rituals. But now such restrictions are not strictly maintained. In Jamespur, a woman (namely Bhadrabati) is found reading out *punthi,* the sacred literature describing the story of *Bonbibi*, on the very occasion of the annual worship of the deity. The recitation of *punthi* is the compulsory part of the worship. Earlier it was compulsorily a man's job. Ideally the *punthi* is to be read out by a Muslim but nowadays they are not very particular on the point. Any person from the worshippers can read it out.

In the folk tradition *Dakshin Rai*, the tiger king is no less an important character. He is worshipped along with *Bonbibi*. In *Bonbibi Jahuranama Dakshin Rai* is portrayed as an anti-hero who in disguise of a tiger goes for human prey. But in *Raimangal*, a story focusing the heroism of *Dakshin Rai*, he is described as a hero. This difference is clearly visible in the folk tradition of the sub-cultural zones on the eastern and western sides of the river Matla. In the *Bonbibi* cult zone *Dakshin Rai* is a malevolent spirit while in *Dakshin Rai* cult zone he is worshipped as the presiding god (*Dakshineswar*). In Dhapdhapi under Baruipur Police station, there is a temple of *Dakshineswar i.e., Dakshin Rai*. Here he is portrayed as a Raja, the king with a gun in his hand and bow and arrows on his back. There is a small wooden figure of the deity dressed as a *shikari*, the hunter, with a gun in his hand. There is no reference of tiger associated with the deity. But in the *Bonbibi* cult zone he is always portrayed in reference with a tiger.

The worshipping place (*than*) for *Dakshin Rai* is erected in seclusion with that of *Bonbibi* and *Shah Jangali*. After worshipping, the forest goers conclude the ritual by saying '*Alla Alla'* (the almighty in Islamic tradition) in the name of *Bonbibi* and by saying *'Hari Hari'* (the almighty in the Hindu pantheon) in the name of *Dakshin Rai*.

It is interesting that *Bonbibi* is worshipped always in association with a set of deities. Apart from *Bonbibi* there are *Dakshin Rai, Shah Jangali* (brother of *Bonbibi*) and *Barkhan Gazi*. The icon compulsorily includes *Dukhe* on the lap of *Bonbibi*. The icon does not include the image of *Ali Madat* but *hajot* (offerings) are compulsorily made in his name. The icon actually includes the theme characters of the myth. *Ali Madap* might have been adapted from some other tradition. In Pakhirala village there is one family who worship *Ali Madap*. The family transcended the deity from their forefather and are continuing the tradition. *Ali Madap* is worshipped for the welfare of the family. He cures diseases, saves the children from the worse, and so on. The people are not sure why *Ali Madap* is offered *puja* in the forest along with *Bonbibi*. There might have been some context in the society but people have forgotten in due course.

Certain taboos are implied on the family, particularly on the wife of the forest goers. Taboos are implied as soon as the team starts for forest and continue till they return. Taboos are important in the folk belief. It is strongly believed that the safety of the team depends on the activities of the women in the family. The wives of the forest goers are prohibited to apply *sindur* (vermilion mark) on the hair parting, to wash clothing with detergent, to comb hair, to light the hearth on Friday, to throw away ashes of the hearth, to visit crematorium, and so on. The taboos resemble that of death pollution or *asouch*. Womenfolk also make ritual prayer to the deity for the safety of the party.

Before the party starts the whole team remains busy in preparation. This is also a joint venture. During the preparation they avoid listening to any discussion that carries any inauspicious message, particularly that of tiger.

The last day of the month of *Paus Sankranti* (Dec-Jan.) according to Bengali calendar year is fixed for annual worship of *Bonbibi*. But because of the preoccupation of the people in the worship of *Bastu devata*, the god meant for homestead on the very day of *Paus Sankranti*, the annual worship of *Bonbibi* has been shifted to the first day of Magh (Jan-Feb). on that day no expedition is made. This very fact again proves that *Bonbibi* is a later entry in the folk tradition. The very fact also shows the dynamism of a folk tradition. New traditions are accommodated in their own way and the change becomes the norm. According to a veteran fisherman the fortnight around the day of annual worship is a dull period for the fishermen as the supply of fish decreases to a great extent in the river.

With full faith on all these beliefs and rituals the people also keep their trust on the alertness and experiences of the teammates. They attach more value to the fellow feeling within the group. They do not submit easily to nature and try to overcome all odds with their intelligence and physical strength. They fight with the tiger with a simple stick in hand, win out the tiger and become local heroes. The credit always goes to *Bonbibi*. Their experiences teach them to be brave. They are to make sound louder than the tiger in direct confrontation. They believe that they can show the bravery only if the deity blesses them. This way the beliefs, rituals, reality and wisdom act in a complex network called the belief system.

Interrelationship between Forest and people

It has already been stated that the relation between forest and people in Sundarban is very old. The old people recollect from their memories that, whenever they wanted wood for posts, for boat or firewood, they just sailed across the river to forest bank and collected the requirements. Sundarban's woods are useful in many ways. As noted by Shib Sankar Mitra (1988: Preface) the trunk of *Sundri* tree was the most preferred item for making boat. The bark of *Garan* tree was useful for dyeing of hide, the oil from the *Garan* tree was specially wanted for polishing of the famous earthen dolls of Bengal; *Genyo* wood was specially used for making of the musical instrument like *dholok* and *tabla*.

Hunter observed (rep.1998: 21) 'the wood cutters of Sundarban appear to have had all along a prescriptive right to fell the forest timber and no government revenue is realized from them'. He (Ibid: 19) further observed that 'the navigable channels of Sundarban was public and no revenue for it was collected on behalf of the government. Turtles, tortoise, crabs and oysters were other creatures of human target.' He cited the then commissioner's report (Ibid: 20-21) that there was no means of ascertaining what proportion of the Sundarban population live by fishing, boating or other industries. All the poorer classes, however, employ themselves in fishing and as boatmen or woodcutters, as a subsidiary means of livelihood in addition to cultivation. All these earlier accounts show that the people have a close association with forest since long, even earlier to British period. The legends and tales also speak of a close relation between man and nature symbolically. This age long interaction between man and forest, developed a unique cultural tradition intermingling man, forest, animals and supernatural beliefs into a complex set of interaction, thereby forming a structure.

The habitat (forest) offers the potential economic resource, which in turn creates circumstances to develop certain social norms, prejudices and traditions. All together it comes to a close-knit interrelationship, thereby forming a definite relationship within the network (Fig. 2). Forest plays a crucial role in maintaining the relationship between the superstructures. Supernatural beliefs and practices, oral traditions and myths in turn regulate their social customs and behaviour, religious notions and value system as a whole. This again forms a domain in specific structural terms.

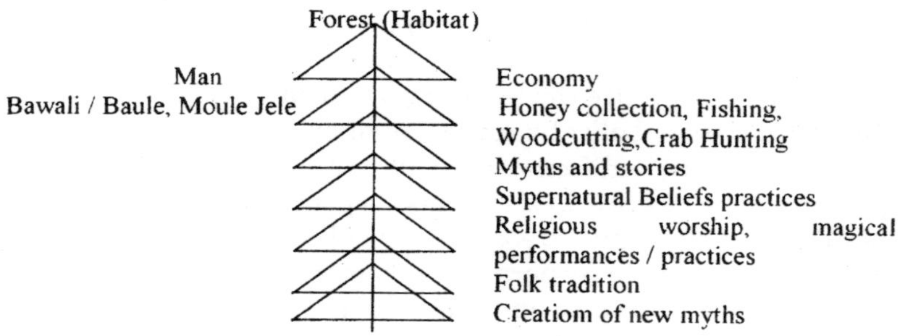

Fig. 2.

Symbolism of forest in folk tradition

In the cultural tradition of Sundarban forest is represented mystically. The myths, stories, beliefs, rituals-all revolve around the mysticism of forest. Myths occupy the central position in cultural tradition regulating the belief system that in turn regulates the ritual practices, thereby forms a domain in structural terms. Elements in a ritual are symbolical to the ideas that act behind. In a folk tradition rituals represent the ideas that have relevance to their way of life, and the myth symbolically represents the interactions between the various facets of life.

In Sundarban the myth of *Bonbibi* and *Dakshin Rai* occupies the central position. The rituals, beliefs and practices mainly revolve around this myth. There is another myth prevalent in this area and that is of *Gazi, Kalu, Champavati* (see Appendix). The later one is gradually fading out in the folk-way of life as it has no direct relevance to the social life. It's importance lies in the reconstruction of the obscure part of the history of Sundarban. Both the myths are characteristically represented through motif.

Both the myths, if interpreted rationally, carry the memories of the forgotten history of some ancient past when the land was invaded by the Islamic preachers who overpowered the then rules of non-Islamic faith, presumably of Hindu pantheon. These also depict a process of transformation of the warriors into the divine characters or otherwise divinity had been imposed on the real characters.

The myth of *Gazi Kalu Champavati* gives us an idea of how these *Gazis* owned the heart of the common people and earned the status of a *pir* (pir are the human reincarnation of the divinity). These *Gazis* (symbolically represented by *Zinda Gazi* and *Kalu Gazi*) usually hailed from the royal families, were actually the warriors. They were a wealthy class of people. Sometimes due to family struggle or due to some other political reasons they took refuge in the remote corners of the country where it was easy for them to mobilize the poor and backward people in their own favour by investing wealth for the welfare of the poor. This way they used to establish their ruling over the small remote areas. The events described in the myth that *Gazi Kalu* established *nagar* (the town), magically made rains of gold coins are nothing but the symbolical expression of the fact as discussed above. They might have used the wealth they carried with them in favour of poor people. With such welfare activities they earned the loyalty of those poor people and became a *pir* for them. With this huge force of local people (symbolically bagh *i.e.,* tiger, kumir *i.e.,* crocodiles, pari *i.e.,* fairies) Gazi and Kalu dared to attack the kingdom of Sonagaon and of Mukut raja, defeated them and converted them into Islam. They applied all unfair means to win out the local kings. As a gesture of submission *Mukut Raja* gave his daughter *Champabati* in marriage to *Gazi.* The myth of *Gazi Kalu Champabati* speaks more about the struggle of power between the Islamic propagators and the local kings than about the crisis and life struggle of the commoners. In other words the story failed to relate the symbols with the real life, as has been done successfully in the other story *i.e.,* of *Bonbibi.* That might be the reason the myth of *Gazi Kalu Champabati* lost its ground in the folk tradition.

The myth of Bonbibi is pivotal in the folk tradition of Sundarban. Like the story of *Gazi Kalu Champabati*, the middle part of the story portrayed the victory of Islamic propagators (*Bonbibi* and *Shah Jangali*) over the ruling of Hindu pantheon. But the success of the story lies in transforming a warrior to divinity and in relating the events with the need of the society. The myth actually earned its base in the folk tradition through the story of *Dukhe* and *Dhona maule. Dukhe* symbolizes the helpless humanity before the hardship of life and *Dhona* symbolizes the exploitation by power. He represents the affluent class to bring out the contrast of the society. This contrast is visible through the myth. Thus, we find *Dukhe* (poor) and *Dhona maule* (rich), the life through the mercy of *Bonbibi* in contrast to death (the tiger king), the extreme helplessness on one hand and the saviour on the other hand. The sacrifice of *Dukhe* to *Dakshin Rai* might be a clue of the old magical practice of human sacrifice to appease the deity. *Dakshin Rai* might be the dominant deity of those days as *Bonbibi* was yet to be introduced in the tradition. *Bonbibi* appeared before *Dukhe* as a saviour in the form of his mother and she not only saved *Dukhe* but also blessed him with wealth and prosperity- an ideal presentation to be an all powerful mother goddess.

The first part of the story dealt with the birth of *Bonbibi* and her twin brother *Shah Jangali.* It is intimated in the story that the birth of twin deities was pre decided in the heaven. She was born in the midst of the forest, deserted by her mother, brought up by wild deer because she was to serve for the greater cause of Islam. The motifs here are common in folk tradition. *Bonbibi* came to *Atharobhati* with a definite goal of propagating Islam. The myth might have been a symbolical representation of the past events when this forest tract was ruled over by the Brahman king. *Dakshin Rai* was descended from *Dandabaksha muni* the ruler of *Atharobhati*. The word *muni* is an adjective used for the Brahman in earlier days. The copper plate inscription of Adilpur (Bakarganj district, (Now Bangladesh) dated 1136 Sambat *i.e.,* 1079 A.D. recorded the grant of three villages by Madhava Sena, the king of Bengal, to a Brahman, who, with landlord rights, received the power of punishing the Chandabhandas,the race that lived in the forest (c.f. Hunter:rep.1998 : 103). This might not be unprecedented one. Sundarban might have been under Brahmin ruler for long and the poor subjects might be looked down upon. Thus, the Islamic preachers and the *Gazis* were heartily welcomed by the local people. Both the myths portrayed the same sentiment of the people.

The myth of Bonbibi, being placed at the center of the complex network of interaction between man, habitat and economy, act as a regulating factor for the belief system in the folk tradition. *Bonbibi* is often treated as forest deity. But the myth of *Bonbibi* never posed her as a deity of the forest. She is in fact the saviour from all possible evils in the forest. She brings the life back to the settlement. The story of *Bonbibi* relates her to the reclamation of forest and to human settlement. For worshipping of her, even into the forest, a hut like structure is erected compulsorily. The construction of hut for the deity symbolizes her linkage with the settled life, not with that of forest. In due course the focus might be shifted to forest economy but the folk tradition symbolically carrying forward the forgotten importance of the deity. In contrast no such structure is required for *Dakshin Rai* as he is associated with the forest and beasts throughout the story. This way the folk tradition carries the remnants of old practices, ideas, beliefs of social relevance through symbols.

According to Sujit Kanti Sur, the researcher on Sundarban, Bonbibi denotes the sister goddess (bon in Bengali means sister). She is referred to as sister goddess in relation to *Shah Jangali,* the twin brother of *Bobibi*. Sur observed that two distinct spellings are maintained in *Bonbibi Jahuranama*, the scripted text of the myth, one is the *bon*, meaning sister and the other is *ban*, meaning forest.

Bonbibi is often confused to *Bondevi*. But the concepts of two deities are distinctly different. *Bonbibi* emerged as an independent concept to that of *Bondevi*. The Rajbanshis of Gosaba clearly recalled that *Bonbibi* was worshipped only in Bholanathpur in the Khulna district. According to Soumen Banerjee of Pakhirala, the forest goers of the Khulna district worshipped Bondevi. According to him *Bondevi* is altogether a different concept of that of *Bonbibi*. *Bonbibi* is hailed from the Islamic tradition while Bondevi is of Hindu tradition. Bondevi typically sits on a tiger and she is having four hands and three eyes, a typical expression of Vedic deity. The *Bonbibi* on the other hand resembles a humanly appearance. Sometimes in the icon she is shown as sitting on a tiger but ideally she sits on a carriage holding *Dukhe* on her lap. *Bondevi* is not associated with any character like *Dukhe*. It is learnt that there was only one person who preferred to make the icon of *Bonbibi* sitting on the tiger. But it is not the usual pattern. In the icon of Bonbibi the tiger god is portrayed as anti-hero. But in the icon of *Bondevi* no such concept is attached. Moreover, Brahman priest is compulsory for *Bondevi* while for *Bonbibi* no priestly service is required from Brahman. It is also informed by Soumen Banerjee that *Bondevi* is still worshipped in Hiranmaypur and Jyotishpur under Basanti Police Station and in Khejurberia in Hasnabad.

The forest goers worship *Bonbibi* on the bank of the river, just at the outskirt of the forest, on the open space. According to Sujit Kanti Sur the bank between the river and the forest symbolizes the boundary between the world of magic and religion. The worship involves the organized behaviour of offerings, salute, and prayers and so on, while magic involves the *mantra* (magic spells) at the individual capacity of the *gunin* or *baule*.

The *mantra* (magic spells) are of four types: *hukumer mantra, guner mantra, paona mantra* and *ekpuruser mantra. Hukumer mantra* compulsorily belongs to a school of *fakir* or *pir*. *Mablej fakir, Alimadap* are some of such schools. *Guner mantra* is a collection of *mantra* from various sources on individual interest. The *mantra* achieved directly from *Bonbibi* is called *paona mantra*.

Ekpuruser mantra is believed to cause the destruction of family by one generation. The very name *ekpurush* implies one generation. According to Phani Gayen, a *gunin* by profession, *hokum* implies a special mental strength *i.e., jor* to overpower the tiger. In the name of *Bonbibi* the *gunin* exerts *hukum* (order) on the tiger and it actually works in the strange situation of forest. Thus, the success of a *hukumer mantra* actually depends on the bravery of the *gunin*. According to Dibya Mandal,a veteran fisherman, it is the mental strength of the *gunin* on which the group depend and it is the solidarity and the bravery of the group on which the *gunin* depends. This is precisely the reality of the situation. Surhid Sardar of Dulki viewed that *gunin* can do nothing unless he gains moral support from the team to face all odds. *Mantra* is nothing but is instrumental to evoke the inner strength of the forest goers. These *mantras* are of folk tradition and expressed in simple colloquial language. Usage of slang is another dimension of *mantra*. The use of slang is common in magical practice. Slang is associated with the practice of driving out the evil spirit. *Mantra* symbolizes man's inherent fear to nature and his intention to change the uncontrollable into controllable.

The myths and symbols have immense influence on the forest goers. Often the characters of the myth are identified with the real life. The case of Haripada Mandal of Pakhirala may be cited here. Haripada started his carrier as forest goer in early childhood (when he was about 10 years of age) because of poverty and pressing need of life. His mother did not want him to go to forest but was helpless. Then Bonbibi appeared in his dream to prevent him from going to forest. She said "Dukhe, don't go to forest. The life in forest is tough". In the dream Haripada replied "Oh mother, you give me shelter in your lap, please save me from all evils. I have no way but to go to forest". The goddess agreed and she blessed Haripada with two words. Haripada took up fishing as occupation depending on those two words. He got married and lived happily. After long days Bonbibi again appeared in his dream and ordered him to leave forest forever. Haripada obeyed the order and stopped going to forest. If the story is analyzed we find the same motifs as in the myth. Haripada was addressed as *Dukhe* because Haripada wanted himself to see in the place of *Dukhe*. He identifies himself with the helpless creature of *Dukhe*. This case is the ideal example of flexibility between the myth and reality. The characters of the myth are known faces in their real life as they can easily identify any of the characters in their social surroundings. Thus, Haripada easily identifies himself with *Dukhe*, the helpless poor child who had no option but to go to forest. The myth and symbols are closely intermixed with their own experiences of life and boundary between the myth and reality abolished. Facing the worst of life in day-to-days life placed these people with the status of *Dukhe*. Thus, we witness the breakdown of Ajhar Gazi into tears while narrating the story of *Dukhe* in relation to *Bonbibi Jahuranama*. His tears speak the success of the symbol of *Dukhe* in the myth of *Bonbibi*. This way myth and symbols mix up with the real life and earn their relevance in folk tradition.

Conclusion

The very environment of Sundarban demands a social solidarity between castes and communities. This is true in every sphere of life, starting from economy to religion. Responding positively to this basic requirement of cohesiveness, the castes and communities maintain a mutual co-existence and tolerance to each other's traditions. Thus, the typical '*Bhasan gaan*' (folk songs sung to appease goddess *Manasa*, the deity of snakes) of Jessor-Khulna region, the '*Jagoran*' (*Sitala Jagoran*) songs of Rajbanshis or the *Tusu* songs of Chotanagpur plateau find a place in the folk tradition of new social settings. The unique environmental *niche* developed a unique cultural base, shared by the people from all regions in common; thereby formed a distinct folk-tradition.

The age long association resulted into assimilation into cultural practices, beliefs and rituals. Rituals in turn create social solidarity. The traditional notion of community here cut across the ethnic boundaries in rituals. Thus, *Bonbibi*, being hailed from Islamic tradition earned trust of forest goers irrespective of caste and creed. Thus, we find Uttara Biswas, a middle aged woman from Rajbanshi caste keeps *roja* (Islamic word for ritual fasting) on the day of worship of *Bonbibi*. She adopted this word from her mother-in-law. We also find the practice of making *hajot* (Islamic word for making offerings) to Bonbibi or to conclude the worship by saying *Alla* in the name of Bonbibi and by saying *Hari* in the name of *Dakshin Rai* irrespective of religious affiliation. It also portrays an ideal example of cultural assimilation.

Emperical data reveal that the wide popularity of Bonbibi is a new addition to the folk tradition. In their original homeland of Khulna dependency on forest was limited to a small group of woodcutters only. The people who migrated to this land were mainly agriculturists and they came with the hope of a more flourishing agrarian economy. But the land proved less fertile and less profitable due to scarcity of sweet water. Gradually the forest based economy gained popularity. This new economic resource demanded new sets of beliefs and rituals, new ideas and ideals. Thus, emerges the concept of *Bonbibi*. The old concept of *Bondevi* might not prove enough to handle the situation. With the concept of *Bonbibi* there emerged the unique folk tradition with unique system of beliefs and practices promoting a social solidarity across the boundary of stratification and diversification of social groups.

A folk tradition is not static. With the dynamism of society tradition also changes. The change may not be visible at once. The old practices are fading out and new practices are making their ground. The agriculturists, who migrated to this land, brought with them the tradition of *bhasan gaan, jagoran gaan*. Nowadays these traditions do not find relevance in society. Rupa Mandal, a Rajbanshi by caste, submitted that young girls do not show interest towards such tradition. Another instance may be cited here is that of *Hari-Sanyasi*, the twin deity once largely worshipped by the Rajbanshis in the early years of their immigration is no more surviving among them. This significance of the deity is even forgotten. Instead they are worshiping *Bonobibi*, which is most relevant in their present way of life. Interestingly *Bonbibi* is also given salute in the *bhasan gaan* along with all other deities. This shows the dynamism of the folk tradition as well as the dominance of the cult. The role of *Bonbibi* is also undergoing change. Earlier she was only a saviour deity. Nowadays she is also a wealth giving deity. Thus, earlier the prayer was only of safe return; now prayer is also made for good catch.

Let us have a cross-sectional view of the situation, Dependency on forest increases because, firstly agriculture is not profitable because of its mono-crop pattern. As the rivers carry saline water, agriculture solely depends on rainwater. The level of ground water, although is high (according to a scientist on the particular field) but pumping or lifting of ground water is discouraged because of the salinity in the surrounding river water. The high level of water in the earth, resists the absorption of saline water, thereby protecting the land from salinity. The farmers are encouraged to raise cash crop by the local authorities but the project often fails due to non-availability of appropriate rate. Secondly, the mono-crop pattern reduces the demand of labour. In agriculture labour is required mainly during the transplanting and harvesting operations. Because of single crop pattern, work is available only twice in a year. Hence people migrate to other regions for want of job. Thirdly, after abolition of the zamindari system in 1953, land was distributed amond the share croppers and landless resulting into small land holdings per family. This further reduces the demand of agricultural labour. No less important is the fact that the yields often fell insufficient to run a family. Another important factor to be considered in this regard is that land holdings per family are becoming smaller due to sharing of lands in due course.

Against this backdrop scenario the forest resource is abundant; is accessible to all; it secures an immediate return in cash; it is in ready access. Moreover, it assures one's bare necessities.

This remarkable contrast between the two economic resources attracting more and more people towards forest based economy. The situation is further worsened because of the backwardness of the area considering the civic amenities, absence of any substitute economic provision to accommodate this large population.

With the growing rush to forest, accidents also occur frequently. This very fact now put the cult of *Bonbibi* in question on many occasions. Baidyanath Biswas, a malo by caste and veteran fisherman stopped worshipping *Bonbibi* as he found that more and more people are falling pray to tiger. Debaranjan Mandal of Emilibari village expressed doubt on the effectiveness of *mantra*. Sudhir Krishna Mandal of Satjelia was of the same opinion but at the same time he admitted that in an extremely uncertain situation they have no option but to keep faith on *mantra*. The faith on the *gunin* is gradually weakening as a number of *gunin* fell prey to tiger. The most famous *gunin* were Girish Baidya and Amir Gazi, both of who were killed by the tiger. But there is the other face of the reality. According to Niranjan Haldar of Malopara, Shyamapada Mistri of Emilibari, Madhai Maity of Dulki *mantra* works, otherwise no fisherman could survive. They believe that *mantras* are the source of mental strength.

The knowledge about the forest since generations developed precise adaptability among the people and this in turn strengthened the bond between the people and the forest. Forest not only embedded deep into their cultural esteem but also assures a livelihood. But one should not overlook the severe poverty and helplessness that push them towards the forest even when their very life is uncertain there. One should value the human thrust for survival in an extremely unfavourable habitat and the bravery of the people in fighting out all odds in life.

The long association with the forest made them equipped with thorough knowledge of the nature. Thus, they know the interrelationship between tides and availability of fish in the river, they can follow a bee deepest into the forest, they can easily make out where the honey is stored in a bee-hive, they can relate the flowering of a particular tree with the quality of honey available, they can make out the season of *Hilsa* fish from the blue colour of the river water or that of *Parse* fish from the typical gray colour of the water. Forest is sacred to them, the *mal* of *Bonbibi*, where she is the saviour, she is the wealth giving deity. Forest is their promise of life. Therefore it is not an easy task to alienate these people from forest, which is not just a piece of land for them, they are associated with the forest through their beliefs, rituals, practices, myths, symbols, precisely by their distinct cultural tradition.

APPENDIX-I

The Myth of *Bonbibi*-A Summary

Bonbibi and *Shah Jangali* were divine souls. *Alla*, the almighty, wished to send them to Earth to take hold of atharobhati, the estate of *Rai* and *Gazi*. They were instructed to take birth in human body by Gulalbibi, the wife of Berahim.

Berahim, a *Faqir* so named, lived in Mecca. He was married to Phulbibi in his youth. They had no issue and were sad for that. Once Berahim received the message from *Alla* that he would be blessed with child by his second marriage only. Berahim got married again to Golalbibi, a fourteen years old daughter of *Faqir* Jalil Shah, with the permission of the first wife. But Phulbibi was not happy with the second marriage of Berahim. Thus when Golalbibi was pregnant by ninth months, she compelled Berahim to desert Golalbibi into the forest.

In the forest Golalbibi found her-self helpless. So she started praying to *Alla*. With the mercy of *Alla* she was taken care of by the divine souls. She gave birth to twin babies-*Bonbibi*, the female child and *Shah Jangali*, the male child. After the birth of the child the divine souls disappeared. Realizing her helplessness to bring up twins without any support in the midst of the forest, Golalbibi picked up the male baby leaving *Bonbibi* behind. Again with the wish of *Alla*, *Bonbibi* was saved and brought up by the divine souls and the wild deer.

This way long seven years have passed Berahim could not bear with her guilty conscience any more and appeared in the forest in search of Golalbibi. He could find out his deserted wife and could managed to convince her to return home. While the couple was on their way to home along with *Shah Jangali*, *Bonbibi* saw them. She appealed to *Shah Jangali* not to go home but to stay back with her for the sake of the holy task they were sent for. *Shah Jangali* and *Bonbibi* reunited and stayed back into the forest. They consoled the parents and send them to their home.

Then *Bonbibi* and *Shah Jangali* reached Madina and got their mission sanctified from the progeny of Hassan. Then they made their way to atharobhati. From Madina they came to Hindusthan. After crossing the river Ganga, they came to the area of one Bhangar Shah who, after being aware of the holy reason of the arrival of the twins showed a friendly gesture to them. From Bhangar Shah they came to know that they already have reached to their destination of atharobhati, the estate of Dakshin Rai. Bhangar Shah also informed them that mom (wax), madhu (honey), khari-jari (woods) of the forest were under the control of Dakshin Rai. *Bonbibi* decided to announce a war against Dakshin Rai. With this idea *Shah Jangali* called ajan (the Islamic way of calling the people for a Prayer). Hearing this call Dakshin Rai got annoyed. He sent his messenger to enquire who had dared to break the peace of his land? The messenger returned with the news of the arrival of *Bonbibi* and *Shah Jangali*. Dakshin Rai announced a war against the Yaban (Muslims) and sent his force to drive them out. But his force retreated with the divine power of *Bonbibi*. Then Dakshin Rai himself came forward on the war. But her mother Narayani held him back for the reason that it does not look nice for a king to fight against a female. So it is she who appeared in war against *Bonbibi*. All the divine power backed *Bonbibi* in war by the mercy of *Alla*. Narayani got a defeat and surrendered to *Bonbibi*. *Bonbibi* owned the right over the Atharobhati. Then *Bonbibi* distributed Atharobhati among the local leaders. One bhati namely Kendokhali was given to Dakshin Rai and kept Bhurkunda under her direct control. She reclaimed the forest and promoted human settlement. This way she established her rule over Atharobhati.

The Story of Dhonai Maule and Dukhe

In the month of Chaitra honey is formed. People reached to the forest for honey collection. Dhonai and Monai, the two brothers, who lived in Barijhati, used to live on mohal (the forest). They were rich. But Dhonai wanted to earn more money and decided to go to Badabon for honey and wax. Monai tried to prevent him because of the fearsome tiger of the forest but Dhonai was determined. He organized seven boats and the sailors, who can accompany him. While the team was about to leave Dhonai felt the shortage of another person. He immediately remembered Dukhe, a poor boy who used live on cattle herding. Dhonai approached Dukhe to accompany him to mohal Dukhe's mother did not agree to send him to the forest, as he was the only son of the poor widow. Dhonai insisted Dukhe over and again with the offer of not only giving him money but also getting him married. Lastly Dukhe agreed upon the proposal and convinced his mother. Dukhe came with Dhona. Dukhe's mother advised him to pray to *Bonbibi* in any distress in the forest. The team entered into the forest but failed to trace out any resource of honey. Dhonai got upset. In this situation he received the message from Dakshin Rai, the king of the forest, in dream that he would get honey and wax only if he sacrifices Dukhe to Dakshin Rai. Dhona begged the life of Dukhe, even he offered his own life but the tiger king was very particular about the young boy as his feast. Dhona agreed. It was then decided to offer Dukhe at Kendokhali. Dhona was blessed with enough honey and wax; even he was advised by the tiger king to collect wax more as wax is costlier than honey. Dukhe somehow came to know his fate. He started crying. At this crisis he remembered the advice of his mother and started praying to *Bonbibi* for a rescue. His prayer reached to *Bonbibi*. *Bonbibi* rushed to him in the disguise of his mother. Dukhe became surprised to see his mother in the midst of the forest but by simple logic he could understand that it is the saviour goddess, who appeared in the form of his mother. Seeing the innocent boy, *Bonbibi* disclosed her identity and consoled him. While returning, Dhona, as promised, stopped at Kendokhali and asked Dukhe to get down on the riverbank to fetch dry woods. Dukhe refused to get down but was compelled by the sailors to do so. As soon as Dukhe reached the beach, Dhona asked his people to set the boats on sailing. They left the place with honey and wax, leaving Dukhe behind. Dukhe find himself in front of the tiger king. He remembered *Bonbibi* at once. Sitting in Bhurkunda, *Bonbibi* heard the anxious prayer of Dukhe and in no time appeared before him along with her brother *Shah Jangali*. *Bonbibi* found Dukhe lying unconscious before the tiger king. She took the young boy in her lap while

Shah Jangali drove out the tiger king from the periphery. Raimoni fled across the river but *Shah Jangali* ran after him. The tiger king set his crocodile force upon *Shah Jangali* but they were defeated very badly. Dakhin Rai came running to Barkhan Gazi. After a while *Shah Jangali* came chasing him. Seeing the situation Barkhan Gazi mediated between the defeated king and *Shah Jangali*. *Shah Jangali* brought them to *Bonbibi*. Gazi reconciled *Bonbibi* with great effort. Then *Bonbibi* ordered Dakhin Rai and Gazi to make friendship with Dukhe and as a token of friendly gesture they should compensate him with wealth. Both Dakshin Rai and Gazi agreed upon the proposal and promised to give him immense wealth.

In the mean time Dhonai returned to the village and informed Dukhe's mother that her son fell prey to tiger. With extreme shock the old lady became blind and deaf. She understood that Dukhe did not pray to *Bonbibi* and thus the tiger could succeed to take him. With the tears of the bereaved mother, the seat of *Bonbibi* reeled under the shock. She immediately sent Dukhe back to his mother with Seko, the pet crocodile of her. In no time Dukhe returned to his village and found his mother in dying stage. Again with the mercy of *Bonbibi* Dukhe's mother got recovered; she also got her eyesight back and got cured of deafness. The mother and the son reunited and became happy. Then Dukhe's mother asked him to offer hajot to the deity by begging alms from seven villages. Dukhe did so and propagate the fame of *Bonbibi*. Dakshin Rai and Barkhan Gazi also gave him immense wealth as promised. With this Dukhe became a rich man. Dakshin Rai provided wood to make house and with the order of *Bonbibi*, one Yadurai constructed a big palace for Dukhe. With the mercy of *Bonbibi*, Dukhe became the Chaudhury (the local landlord or the ruler) of the area. After seeing the wealth and power of Dukhe and also remembering the past events Dhona became afraid. He begged the mercy of *Bonbibi*. The kind mother advised him to give his daughter in marriage to Dukhe. Dhona agreed. Thus Dukhe got married to Dhona's daughter. On this happy occasion Dhukhe remembered *Bonbibi*. She appeared in the disguise of a white mosquito and blessed the couple. Dukhe became a good ruler and lived happily.

APPENDIX-II

The Story of Gazi Kalu Champabati-A Summary

Gazi was the son of Shah Sekendar of Bairat Nagar. From his childhood, Gazi was religious minded and denounce to be succeeded to throne. Being angry, Shah Sekendar punished Gazi to death but Gazi could overcome all the distress by the mercy of *Alla*, the almighty god. Kalu was the adopted son of Shah Sekendar and like Gazi he was also of religious minded. Both of them left home one night to propagate the name of *Alla* and his charisma. They traveled a lot and wherever they had gone, made the people to accept Islam and to become their disciple. They came to Sundarban and stayed there for seven years. All the tigers, crocodiles and fairies submitted their loyalty to them. From Sundarban they moved to Chhapai Nagar and compelled the king Sriram to embrace Islam. They moved further on to Brahmanya Nagar, defeated the king namely Mukut Raja with tricks and compelled the King to give his daughter Champavati in marriage to Gazi. The couple was living happily but Kalu wanted Gazi to leave his wife behind and to devote himself for the cause of *Alla*. Gazi was convinced and one night they took chance to escape from the palace of the King. But Champavati could smell the plan and Gazi could not avoid her. All three left home but in the midway Champavati was deserted by Gazi. After long way of Gazi and Champavati reunited. The story ends with happy reunion of all the family members of Gazi and Kalu.

The story tells the details of love affair between Gazi and Champavati and also narrates in minute detail the tricks played by Gazi and Kalu to compel the Hindu kings to formally convert to Islam.

References

Census 1951, 1953, *The Tribes and Castes of West Bengal,* West Bengal Govt. Press, West Bengal.

Hunter, W.W., 1998, *A Statistical Account of Bengal,* Vol. I, Part-II, Statistical Accounts of The Sundarbans, West Bengal District Gazetteers, Calcutta, (Original Print 1876).

Jalil, A.F.M. Abdul, 2000, *The history of Sundarban* (in Bengali), Naya Udyog, Kolkata.

Kabikankan Mukundaram, 1975, *Chandimangal,* Calcutta University Press, (Reprint).

Marhum Munshi Mohammad Khater Saheb, 1401(Beng. Calendar year) *Bonbibi Jahuranama* (in Bengali), Gayosia Library, Kolkata.

Mitra, Shib Shankar, 1988, *Sundarban Samagra* (in Bengali), Ananda Publishers Pvt. Ltd., Kolkata.

Munshi Abdur, Rahim Saheb, *Gazi Kalu O Champabati*, Kanyar Punthi (in Bengali), Gayosia Library, Kolkata. Year of Publication illegible.

O' Mally, L.S.S., 1914, *Khulna District Gazetteers.*

Risley, H.H., 1891, *Tribes and Castes of Bengal*, I (cited from Census 1951, The Tribes and castes oif West Bengal, West Bengal Govt. Press, West Bengal, 1953).

Sur, Sujit Kanti, 1982, '*Banglar Prachintama Lokpuran-Manasamangal*' (Bengali) In *Aritra*, 55-90.

4

Contact and Conflict in the Sundarbans Biosphere Reserve

Krishna Basu

Introduction

The relationship between human being and physical environment has been in existence for a very long time. The mode of economic value of the physical environment and the pattern of utilization by the human being is different. Since the early historical days, the kings used the forests as a pleasure spot and hunting place. 'In the middle ages in Europe, industrial worker's influenced the emergence of new attitudes to work and life, which were to alter their approach to leisure, which was undoubtedly the early twentieth century development of paid holidays' (Urry, 1990). Since the development of colonization we learn and lesson from it, that the process of utilization pattern is resulting in degrading or deforestation of ecology, and overnight build up of industry without respect of physical environment. In the context of Sundarbans delta, it offers outstanding scenic beauty and great diversity of plant and animal life, presence of mangrove forest. Undoubtedly, the Biosphere is magnificent and unique. The Sundarbans delta has been declared as a "World Heritage Site" in the year of 1987 and "Biosphere Reserve" in the year of 2001.

On account of empirical view, in the name of development of higher perspective, we find both the negative and positive aspects as two sides of the same coin. In course of colonization of the delta, relationship between the people's behavior and use of forest resource became unprecedented. Today, the people were settled in areas adjacent to the Biosphere Reserve. As a result, many unpleasant encounters took place between man and animal or man and authority. All these resulted in vengeance and loss of live on both sides. On the other hand, The Government of India as well as Government of West Bengal, Tourism Department, has taken a policy on Eco-tourism, the objective of which was to develop the economic condition of the local people as well as to develop the national economy. Today the tourism industry is a global phenomenon all over the world. In India, nowadays tourism is also considered as a promising industry.

Sundarbans is the largest delta in the world, having rich biodiversity of both terrestrial and aquatic resources. The area consists of vast tracts of forest and swamps. The forest happens to be the only mangrove tiger land and rich biodiversity land in India. This tiger reserve and its bio-diversity is considered most important in the area as well a "World Heritage Site". But a sizeable number of people came into contact of the Biosphere resources and utilized them for their own consumption as well as commercial purpose. On the other hand, the tourism industry has been developed in the biosphere and provided all infrastructural facilities for the tourists at the same time the local people fulfill the tourist needs from the Biosphere, it eventually became direct or indirect threat to the biosphere Reserve.

Keeping in mind the potential strength of the biosphere, if we hamper the environment, extinction is the most serious issue because it is irreversible. Extinction is happening at unnatural rates in many parts of the globe wherever plant and animal species in small range come into conflict with human demands. Bio-diversity is crowded out as mass consumption and puts pressure on space and natural resources. In the context of Sundarban Biosphere is not an exceptional field. The study tries to carryout present scenario of the Biosphere, which is under pressure.

The following diagram considers, since early 18th century, the British came to the area for colonization and later on, set up a policy, *i.e.*, clearing the forest and introduce the cultivation in the same clearing land. The British brought tribal communities mainly Oraon, Bhumij, Munda, and others from different parts of country like Chhotanagpur, Mayurbhanj, Ranchi, Purulia, Midnapur and adjoining areas. On the other hand, communities like Mahishya, Poundra Kshatriya, Mahatos, Namosudra, Bagdi, Tantubai, Kamar; Dom and others also came to this region for the said work.

BIOSPHERE IS UNDER PRESSURE

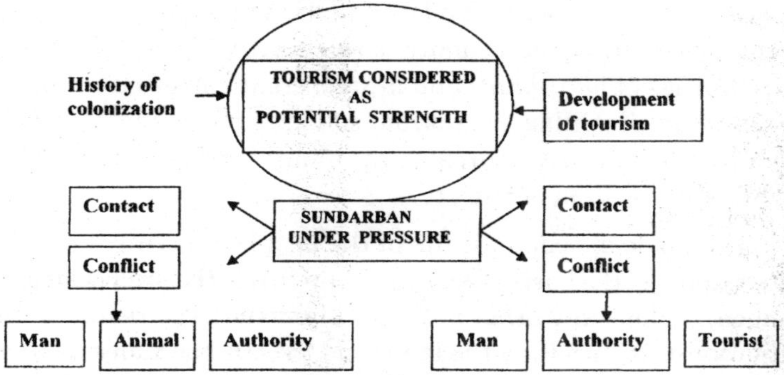

Without potential strength of the area:-
We cannot run tourism industry
We pay for it.

In 1947, due to partition of India and Pakistan, the populations came to this area mainly from Khulna, Jessore and other districts of East Pakistan and settled in the areas adjacent to the Biosphere Reserve.

In 1970, due to Bangladesh war, during the period. There was a sudden influx of population in this area and population pressure over burdened the Sundarban Biosphere.

As a result of the influx of population pressure, a hazard situation was created in the Sundarban area also affecting their economic life. On the basis of, the present scenario of the Sundarban delta lost its "Carrying Capacity" and the "Man and Land" ratio were imbalanced.

The second part deals with the development of tourism in the area, *i.e.,* biosphere reserve and adjacent area. To emphasise how the biosphere reserve is under pressure due to impact of tourism. Tourism department has initiated development policy, for improving local people's economic condition as well as improve the national economy. The authority provides better facilities to the tourist like transport, accommodation, communication, etc, in the Biosphere. The local people also satisfy the tourists by providing honey, different types of fish, crabs, etc., which they collect from the same biosphere reserve. The next focus is the tourist, they also damage the same area by garbage which is generated by the tourists, and dumped into the river which also creates an environmental pollution. During the peak tourist season the activities of *bhat-bhati* and launch create an environmental pollution by oil spillage, smoking, soil erosion, sound pollution, etc. It is also indirectly a threat to the mangrove forest, hampering the breeding system of fish etc.

The mode of economic life of the people is mainly based on cultivation. On the basis of ecological condition, the area is affected by flood, cyclone and other natural calamities. The result is that saline water is engulfing the land, and the land becomes unfit for cultivation for the next few seasons. It requires at least three to four years getting the land free from salinity and it is threat to their economic life. The people have no other way, ultimately they for an search for an alternative economy, *i.e.,* involving themselves on nearby forest resources, both terrestrial and aquatic. This creates a pressure on the biosphere resources. As a result, the human needs turn into conflict situation between man and animal, and between man and authority. It is clear that, the potential strength of the Biosphere requires human contact like wood for fuel and domestic purpose, hental (leaves for thrashing the house, fish collection for livelihood and commercial purpose, honey collection, shark fin, liver of Ari and Koyal fish, etc., from these items they earn more money for their economic life.

Eco-tourism is mainly based on conservation and preservation of the ecology, at the same time to develop the economic condition of the local people without loss of ecological balance. In the context of Sundarban Biosphere Reserve, the people depend on the biosphere resources, from pre and post colonial periods they came into contact with the potential strength of the biosphere. In the field biosphere the term conflict is applied for the people, where they cross the territorial or notification zones that would be offence and counted as conflict situation, by the authorities and by the animal.

Sundarbans are the largest delta in the world. The Indian portion of Sundarbans is bounded by Dampier and Hodges line in the North, Bay of Bengal in the South, Bangladesh in the East and river Hoogly in the West. Out of the total area of 25,500 sq. kms., 9630 sq. kms. lie in India and the rest is in Bangladesh. Out of 9630 sq. kms. in India, 4264 sq. kms. is forest land and the balance 5366 sq. kms. is demarcated as human habitation. The Indian portion comprises a total of 102 islands; out of which 48 islands are forest covered and rest 54 islands have human population.

The Sundarbans form the southern most portion of the Gangetic delta and extend along the sea-face of the Bay of Bengal, from the estuary of the Hoogly on the west to the river Meghna (Bangladesh) on the east. It is through this area the united water of the Ganges and the Brahmaputra find their way to the sea. Sundarbans of Bangladesh and India is the single largest mangrove chunk of the Glove, lying within the latitude 210N-22'031'N and longitude between 88010'E and 92015' E. This area comprises vast tracts of forest and swamps.

From the administrative point of view the forest department is classified in the different categories.

Core Zone-The area is considered as highly protective zone, which is 1692 sq. kms. It is a completely undisturbed natural reserve and wilderness portion. Under this area belong the "Sundarbans Tiger Project" and the "Sundarbans National Park" (1330 sq. kms.) in 1984 and the "Wild Life Sanctuary" (362 sq. kms.). Sundarbans Tiger project consists of various forests blocks and number of compartments, such as Jhilla, Arbesi Khatuajhuji, Chandkhali, Chamta, Matla, Gona Bagmara, Harinbhanga Gosaba, Chotohardi, Mayadwip, Netidhopani, Pirkhali, Panchamukhani. The Forest Department gives permission for fishing and collection of honey to the local people for Jhilla, Arbesi, Khatuajhuri and Chandkhali Forest Blocks and rest of the forest blocks are prohibited areas.

Buffer zone-The buffer zone consists of 2233 sq. kms. of mangrove forests having three sanctuaries viz: Sajnekhali (362 sq. kms.); Lothian Island (38 sq. kms.) and Haliday Island (6 sq. kms.). The buffer zone of the Sundarbans Tiger Reserve, which also includes the Sajnekhali Bird Sanctuary, is used for fishing and honey collection. But felling of 'hental' (Phoenix paludosa) is banned as the hental tree provides the typical tiger habitat.

- *Transition zone*-The area consists of 5705 sq. kms. along northern boundary of the reserve is some portion of reclaimed Sundarbans where the local people extensively practice agriculture and coastal aquaculture.

Historical background

Sundarbans has an important place in Indian history. Even though, the Sundarbans forest is mentioned in the Mahabharata and the Puranas, nothing definite was known about the area until the end of the 15th century when a few details of the area are found to be mentioned in *Manasamangal* of Bipradasa and in the Ain-i-Akbari. The story of Chandsaagar and the account of Todarmal also give indications of the existence of a vast swampy forest to the south and east of present of Kolkata, as one proceeds along the Adiganga towards the Bay of Bengal. The actual ruler of the Sundarbans towards the end of the 16th century was a chief called Pratapaditya, one of the Baro-Bhuinyas. The exact extent of the Sundarban forests of the period, although is not mentioned anywhere, scattered evidence and records available from various sources indicate that the forests extended from the south-eastern portion of Midnapur district of West Bengal to Bakharganj district in East Bengal covering almost the entire portion of 24 Parganas district (West Bengal), Khulna district and Barisal district (East Bengal). Records about Sundarbans, including the portion in East Bengal, show that in 1895, the total area of Sundarbans forests were about 20,000 sq. kms. of which 60 per cent belonged to Bangladesh and 40 per cent of to West Bengal in India. At the time of partition of India and Pakistan, in 1947, about 10,000 sq. kms. of forests were still intact of which 4264 sq. kms. devolved to West Bengal. There is evidence of Mughal, Paul, Sen, and Gupta dynastic rule and lastly that of the British. The British started the history of colonization of this area in the early 18th century. The British in 1770 decided for utilization of the land for agriculture. In 1785 Tilman Henckell distributed land among cultivators known as "Ijarådar" for reclamation. In 1817 the land of Sundarbans was declared as Government property. In 1822-1823 Princep completed the initial survey between the rivers Jamuna (now in Bangladesh) and Hoogly to divide the entire area into small blocks called lots. During 1829-1830 William Dampier surveyed the forested area. In 1830 the British framed a policy to distribute the forest area among the British settlers. In 1872 human habitation came to Sundarbans. In 1878-1879, the British declared the area of 4856 sq. kms. as protected forest land.

In the first half of the nineteenth century, the Sundarbans forest, most of the southern part of Diamond Harbour, Alipur Sadar and Basirhat Subdivisions were being deforested and brought under cultivation. Landless labourers and fishermen from the neighbouring areas of Jessore, Khulna and Nadiya in settled in the area. Most of them belong to the Scheduled Castes like Pods or Paundra Kshatriyas, Jalia and Halia Kaivartas, Namasudras and Muslims. In the second half of the nineteenth century, the source of immigration to Sundarbans portion of the district changed and more and more agricultural laborers and marginal cultivators from Medinipur and other comparatively arid parts of south-western Rarha regions of West Bengal and the plateau regions of south Bihar started migrating to Sundarban. The new immigrants included a fair number of Santal, Oraon, Munda, Kurmi and Kurmi Mahatos and Mahishyas between 1881 and 1911, who settled in the area, on or around 1921.

Sir Daniel Mackinan Hamilton, a Scot National came to Calcutta in the first decade of the twentieth century as the Chief Executive of the Mackinan and Mackenzie Company. This benevolent gentleman desired to set up a model colony for the enlistment of downtrodden people. In 1903, Sir Daniel Hamilton, established his Zamindari over 9,000 acres land purchased comprising two islands namely Gosaba and Rangabelia. In 1920 Hamilton also acquired Satjelia comprising 13,000 acres. He brought people for establishing populace and clearing the forests to make the land cultivable. These people mainly came from Chhotanagpur Plateau such as Mayurbhanj, Midnapur, Purulia,

Ranchi and adjoining areas. Along with these the other non-tribal communities like Mahishya, Poundra Kshastriya/Pod, Mahato, Namasudra, Bagdi, Tantubay, Kamar, Dom and others also came to this region for the land cultivation.

After the partition of India in 1947, a sudden influx of migrants from East Pakistan (now Bangladesh) mainly from the districts of Jessore, Khulna, Barisal and Bakharganj were added to the existing population of Sundarbans. Some migrants belonged to scheduled castes like Rajbanshi, Namosudra, Poundra and others like Malo (fisher folk), Jele; besides other higher Hindu castes. Another mass population influx into this area happened after the 1970-71 Bangladesh war. A huge number of refugees forcibly occupied this area and settled in different parts of the Sundarbans. As a result, population explosion in the Sundarbans Biosphere region, the carrying capacity of the area of land became imbalanced in proportion to the "Man and Land" ratio.

In course of migration, these migrants either were rehabilitated or illegally settled adjacent to the Biosphere Reserve. These people were primarily involved in agricultural activities. On the other hand a large number of them were also engaged as daily wage labourers. Due to various natural phenomena like floods, storms, cyclones, etc, used to result in saline water engulfing the land, thus causing infertility. Due to the need for survival they search alternative source of economy for their socio-economic life. As a result, they devoted their efforts and came into contact in illegal collection of produce from the forests, rampant fishing, collection of "bagda meen", deforestation-felling of timber and wood, poaching of animals and other forest based activities as a means of livelihood.

"This situation has resulted in leading to a severe crisis of ecological imbalance that threatens to swallow the natural bounty of flora and fauna of the Biosphere Reserve at Sundarbans".

Population

The human habitation has increased manifold over the years in the Sundarbans area to become a threat to the Biosphere Reserve.

Population of Sundarban (1872-2001)

Year	*Population*
1872	2,96,045
1881	3,55,512
1891	4,19,818
1901	4,87,377
1911	5,70,878
1921	6,48,654
1931#	7,54,421
1951#	11,59,559
1961	15,32,101
1971*	20,03,097
1981	24,55,365
1991	32,05,528
2001	42,00,000

Campus: FEB-2001 (P1395-1401)
#1931-1951 Population increased due to partition.
*1971 Population increased due to Bangladesh war.

The table shows mainly on population pressure since 1872 to 2001 on three factors. In 1901 during the period the area is undivided Sundarban forest and the population was 4, 87,377. In 1951 during the period, populations increased due to partition of India and Pakistan. In 1971 during this period a huge number of population came into Sundarban area after Bangladesh war. According to 2001 census figure, it is clearly indicated that the area is under population pressure and this situation still continues.

Methodology

Under the project "Cultural Dimension of Tourism in the Sundarbans Biosphere Reserve", field-work was conducted in the month of January, 2005 (1st phase) and January to February 2006 (2nd phase) at Dayapur village and adjacent villages under the Gosaba block.

All the conventional tools and techniques were applied for necessary data collection such as: census schedules, questionnaire method, interview method, case study, observation method and genealogical method.

The total number of household surveyed at Dayapur village is 914. Their economic condition, marital status, education, land holding pattern, etc. In addition, data collected from age-old persons both male and female for their long time experiences related to exploitation of natural resources by group discussion. Data was collected from those who are presently involved in forest produce activities and interviewed few of them both male and female members. Besides, the author also acquired perception of tourism and the ecological condition of the area (past and present) situation, from individuals like youth, teachers, and panchayet members through interview. During the time of field work some interview was taken from neighbouring village, *i.e.,* No.2 Satjelia bazar where the forest produce like (a) fuel wood, (b) timber used for house making and commercial purposes are sold in the open market. In addition, the genealogical method has been followed for collection of data from Dayapur village and neighbouring villages like Jamespur, Pakhirala, Lahiripur and Jelepara which reflected their involvement, from generation to generation, as to how the people are involved in illegal activities with the surrounding environment. Besides the above information and data from Government archives from different offices like the Forest Department and the local police station have also been collected.

The Village

The study area considered at *Dayapur* Village of *Satjelia* Island which is under Gosaba block, district of South 24 Parganas and it consists of 14 numbers of anchal. The island is situated very close to the reserve forest. Besides, tourism infrastructure is available in the island, from where they can easily move to different tourist spots. Three sides of the village are almost surrounded by the Gomor River and at the southern part lies the Sundarban reserve forest.

Dayapur is a multi-ethnic village, where the people of scheduled castes, scheduled tribes, other backward castes and general caste live together. However, this is a Scheduled Caste dominated village mainly by Poundra Kshatriyas from Bangladesh. Beside this, Rajbansi, Malo, Bagdi, Namasudra and others are the other major communities concentrated in this village.

Among the tribals, Bhumij and Munda are the earliest settlers in this region migrated from Ranchi of Jharkhand or Mayurbhanj of Orissa. The people are dependent mainly on cultivation, besides a large number of people are engaged and fulfill their needs from the Biosphere. It is also observed that the number of genealogical data shows a huge number of families are involved in procurement from the forest resources of both terrestrial and aquatic nature from generation to generation.

The total number of households surveyed at Dayapur village is 914. The total population of this village is 4274, out of which 2211 are males, *i.e.,* 51.73 per cent and 2063 are females, *i.e.,* 48.27 per cent. The village is altogether twelve number of paras such as Renuka Nagar colony, Mridha para, Topor Dakshin, Topor Uttar, Binapani, Kastura, Abirkhali, Sankobhanga, Banikhali, Bagedhara, Kumirdoba and Adibasi Majher Para. In addition, some case studies was collected from neighbouring villages like Jamespur, Lahiripur, Pakhirala, Jelepara and village where the villagers depend upon cultivation and also involved in collection of the resources from the Biosphere Reserve. It is observed that the maximum number of inhabitants belongs to Adibasi Majher Para, Renuka nagar colony, Binapani, *i.e.,* 784, 637, and 613 respectively.

Table-1 : Age Group Wise Population Distribution-Dayapur

Age Group	*Male* No.	*Male* %	*Female* No.	*Female* %	*Total* No.	*Total* %	*Cum. Total* %
0-4	169	3.954	182	4.26	351	8.21	8.21
5-9	186	4.352	209	4.89	395	9.24	17.45
10-14	255	5.966	235	5.50	590	11.46	28.92
15-19	226	5.288	208	4.87	434	10.15	39.07
20-24	198	4.633	173	4.05	371	8.68	47.75
25-29	208	4.867	207	4.84	415	9.71	57.46
30-34	155	3.627	129	3.02	284	6.64	64.11
35-39	157	3.673	160	3.74	317	7.42	71.53
40-44	115	2.691	122	2.85	237	5.55	77.07
45-49	165	3.861	137	3.21	302	7.07	84.14
50-54	89	2.082	102	2.39	191	4.47	88.61
55-59	118	2.761	74	1.73	192	4.49	93.10
60-64	60	1.40	58	1.36	118	2.76	95.86
68-69	48	1.123	33	0.77	81	1.0	97.75
70-+	62	1.4651	34	0.80	96	2.25	100.00

Economic life of the people

The economic life of the people is mainly depending on cultivation and nearby surrounding of the Biosphere Reserve both terrestrial and aquatic resources. On the other hand, the small percentage of young generation mainly engage themselves in casual labour, seasonal labours in different parts of India like Andaman, Tripura, Andhra Pradesh, etc. The total cultivable land in Sundarban is 3, 10,562 ha. out of which in 35,041 ha, irrigation facility is available.

The ecological condition of the area is not suitable to cultivate the land due to natural disasters. There is every risk of destruction of the standing crops by saline inundation from the sea. In the past, this saline inundation not only destroyed the season's crops but rendered the land unfit for cultivation for a few seasons. This land needs minimum three years to get, fit for the next crop due to loss of salinity. The situation became direct threat to their economic life.

The socio-economic conditions are alarming, vis-à-vis the basic survival needs of Dayapur and adjacent villagers. The landless people and those who have small occupancy of land is dependant on nature's bounty for basic sustenance and economic gains irrespective of the dangers of being attacked by animals and by the authorities for illegal entry into the forests. That the forest offers an alternative scope for economic returns in the form of collection of produce, fishing and forest products, including medicinal plant products is very evident and the people are taking to it.

From the Table-2, it is clear that maximum numbers of people are engaged in agriculture (438), *i.e.*, 10.2 per cent of the total population. Besides, 286, *i.e.*, 6.7 per cent are involved in daily labour outside, 200 population, *i.e.*, 4.68 per cent are engaged as daily labours, 96 people, *i.e.*, 2.2 per cent are engaged as agricultural labours, 302 population, *i.e.*, 7.07 per cent are engaged as *Bagda meen* collection and 30 people are exclusively engaged in forest resources collection respectively.

Table-2 : Occupational Distribution-Age Wise-Primary Occupation at Da-apur

Type → / Age Group ↓	1	2	3	4	5	6	7	8	9	10	11	12	13	14	15	16	17	18	19	20	21	TOTAL
0--4	0	0	0	0	0	0	0	0	0	0	0	0	0	351	0	0	0	0	0	0	0	351
5--9	0	0	0	0	0	0	0	0	0	0	0	0	338	56	0	0	0	1	0	0	0	395
10--14	1	1	6	0	0	0	0	2	2	0	1	0	455	8	0	0	5	8	0	0	1	490
15--19	8	9	30	0	0	2	0	14	33	0	0	0	251	6	1	0	60	11	1	1	7	434
20--24	32	16	49	2	1	5	0	19	46	1	2	0	54	3	6	1	125	2	0	0	7	371
25--29	40	13	44	2	9	12	0	29	68	3	7	3	16	2	9	0	149	2	1	2	4	415
30--34	37	8	34	4	10	7	0	29	38	1	9	2	0	0	7	1	83	2	3	1	8	284
35--39	58	11	31	2	5	9	0	26	32	1	3	2	0	1	7	2	110	1	3	0	13	317
40--44	40	8	33	3	6	7	0	15	23	2	6	4	0	1	6	2	74	1	0	0	6	237
45--49	46	12	45	4	2	16	0	27	26	4	7	5	0	3	7	4	84	0	1	2	7	302
50--54	29	5	10	4	1	8	0	19	11	0	3	6	0	3	4	1	82	1	1	0	3	191
55--59	62	5	14	9	2	12	0	10	5	1	6	1	0	2	2	0	54	1	1	0	5	192
60--64	29	3	4	1	0	3	5	5	2	1	2	2	0	6	2	0	51	0	0	0	2	118
65--69	31	3	1	0	0	0	2	3	0	0	0	3	0	8	0	1	28	0	0	1	0	81
70--+	25	2	1	0	0	0	4	2	0	1	0	2	0	46	0	1	6	1	0	3	2	96
TOTAL →	438	96	302	31	36	81	11	200	286	15	46	30	1114	496	51	13	911	31	11	10	65	4274
%	10.2	2.2	7.07	0.7	0.8	1.9	0.3	4.68	6.7	0.4	1.1	0.7	26.1	12	1.2	0	21	1	0	0.2	2	100
	1	2	3	4	5	6	7	8	9	10	11	12	13	14	15	16	17	18	19	20	21	

Index of Type of Occupation

Occupation	Type	Occupation	Type	Occupation	Type
Agriculture	1	Daily Labour Outside	9	House Work	17
Agr Labour	2	Village Doctor	10	Maid Servant	18
Bagda	3	Shop keeper	11	Bag Business	19
Teacher	4	Forest items collector	12	Jajmani	20
Tution	5	Student	13	Others	21
Service	6	Unemployed	14		
Pensioner	7	Business	15		
Daily Labour	8	Carpenter	16		

Table-3 : Land Holding Pattern of Dayapur Village

Hamlet(para)	*0-5 bigha*	*5-10 bigha*	*10+ bigha*	*Landless Family*	*Total Families*
Abirkhali	36	2	1	2	41
Bagedhara	16	0	1	2	19
Binapani	70	22	1	45	138
Banikhali	13	0	0	11	24
Kumirdoba	57	3	1	10	71
Majher para	94	17	9	44	164
RNC	54	14	3	56	127
Topor-D	34	12	5	8	59
Topor-U	22	21	7	17	67
Sankobhanga	19	3	0	0	22
Mridhapara	48	10	3	65	126
Kastura	43	6	0	7	56
Total	506	110	31	267	914
%	55.36%	12.04%	3.39%	29.21%	100%

Dayapur village comprises 12 hamlets or paras. The total numbers of families are 914 out of which 647 families have land and 267 families are landless (29.21 per cent of the families are landless). The 647 families having land have been categorized as per their land holding possession. Under 0-5 bigha category, 506 numbers of families (55.36 per cent); under 5-10 bigha category, 110 numbers of families (12.04 per cent); and under 10+ bigha category, 31 numbers of families (3.39 per cent).

In spite of the land holding pattern, the people could not depend on agricultural economy because the area is affected by the natural disaster; and it reflects indirectly to their economic life. As a result, the people search for alternative economy and ultimately depend on their surrounding natural resources, *i.e.*, terrestrial and aquatic in the same biosphere for their survival. It is observed that the people who are landless and small occupants of land holding mainly depend on natural resources. As a result, the human demands increase and the trend becomes more viable for contact and conflict between man and animal, and man and authority respectively.

Nature of collection from the biosphere

The causes of contact to the biosphere by the local people is clear from the pattern of immigration, natural disaster, colonization, tourism; the people are more close to the biosphere resources and fulfill the demands for their survival as well as tourist needs. The causes of human demands break some rules and regulations, which is governed by the local authority for the beneficiaries. But the beneficiaries cross their limitation at the time of collection of both terrestrial and aquatic resources. As a result of this, there arises conflict situation.

The nature of exploitation from forest and river or creek by the people of Dayapur villagers and the adjacent villagers, *i.e.*, Jamespur, Pakhirala, and Lahiripur is a common phenomenon. On the other hand the foreign poachers from Bangladesh, Burma and other places also exploit the natural resources in the same area, *i.e.*, the Biosphere Reserve. The types of resources exploited are given below-

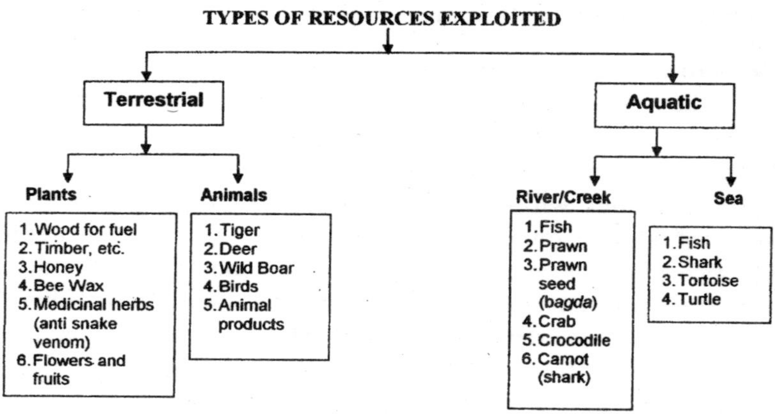

The diagram shows types of resources, both terrestrial and aquatic, of the Biosphere exploited by the local people and foreigners. The resource base is divided into two categories, *viz.* terrestrial and aquatic. The terrestrial resources are divided in to two parts (1) Plant products like honey, medicinal plants, fuel wood and timber, etc. (2) Animal products-tiger, deer, wild pig, birds, etc.

The aquatic resources are also divided into two categories-(1) River or creek resources like different types of fish, crab, prawn and prawn seeds (Bagda), (2) Sea products such as fish, shark, crocodile, tortoise, etc. All the products are used by the people both domestic and commercial. But due to poor transportation and lack of Government co-operative system, they are not getting proper market value.

Fish is considered as a major item of staple food, but due to poor infrastructure facilities local people are exploited by the local agents or the middle men. Another drawback that is faced by the people is that there is no arrangement for the preservation of fish. The types of fishes generally caught by the people are *Bacha, Bele, Bhetki, Boyal, Chanda, Chital, Punti, Gurjaoli, Ilish* (Hilsa), *Chele, Jaba, Bhetki jaba, Selet, Mochan/air, Med, Rupamati, Kan, Payra, Sitapati, Fyasha, Gagra, Chakul*, etc.

Through the above discussion it is indicated that the type of resource utilization or exploitation by the local people and foreigners makes the situation alarming.

Seasonal Variation and Resources

The word cosmology is the cultural worldview which defines a people's understanding of their spiritual universe and gives shape to their myths, rituals, social behaviour and history. A cosmology situates a people within their environment in time and weaves a series of relationships among them with their God, and between other peoples. A people's cosmology is the essence of their identity. It nurtures and sustains life. The people of Sundarban Biosphere Reserve also depend on cosmology which reflects their economic life-like fishing, wood cutting, crab collection, honey collection, etc. It has been observed that the people of Sundarban including the Dayapur villagers are accustomed with the seasonal and lunar system for procurement of the resources from both terrestrial and aquatic resources. The mother cult of "*Banobibi*, and *Dakhin Rai, Shahajangali*, are the central authority of the Biosphere. The people believe that they are the benevolent cults of the Biosphere. Before proceeding for their collection the people appease the cults, for their protection or safety.

Concept of Seasonal Calendar

The local people use age old traditional knowledge which they acquired from generation to generation during the time of collection

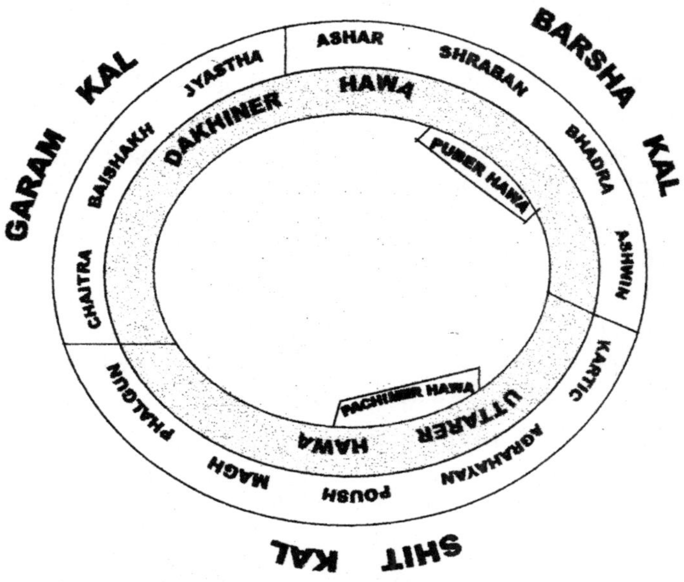

of honey and fish. On the basis of availability of resources the people of Sundarbans divide the seasons as per Bengali Calendar mainly three categories namely *Garam kal* (summer, March to June,) *Barsa kal* (rainy, July to October) and *Shit kal* (winter, November to February) months. As per Bengali calendar *garam kal* is considered from the month of Chaitra to Jaistha, *barsa kal* is considered from the month of Ashar to Kartik and *shit kal* is considered from Agrahayan to Falgun month accordingly. In spite of three seasons the people also procure prawn, crabs and Parse fish throughout the year. During the summer season fishing is prohibited as per the order of the forest authorities. During the time permits are available only for honey collection exclusively in the month of *Chaitra* and *Baishakh*, but females and girl children collect Bagda meen from the river bank round the year, locally known as "*char*".

Rainy season is the best time for fishing. They also collect crabs in small quantity but the rate is very low. During the month of *Bhadra*, Hilsha fish is available but it depends upon weather condition as per their knowledge, water is blackish for a few days. During the months of *Falgun* and *Kartik* women also go to jungle for crab collection. They go to jungle at high tide and are back at home during low tide. They believe that jungle is their second homeland. I have collected some special case studies from women jungle goers who are very much dependent on the nearby surrounding, *i.e.*, natural resources for maintaining their livelihood and having more than thirty years of experience.

Winter season is mainly the best time for collection of crab, prawn and *Bagda meen* (from deep river). Fishes are mainly available which are locally known as *Mochan* or *Air*, *Chakul*, *Rupamoti*, *Chele*, *Parse*, *Chanda*, etc. During this season Hilsha fish is also available for a couple of days, during which time the water colour is bluish, with mixed saliva.

Lunar System and Resource

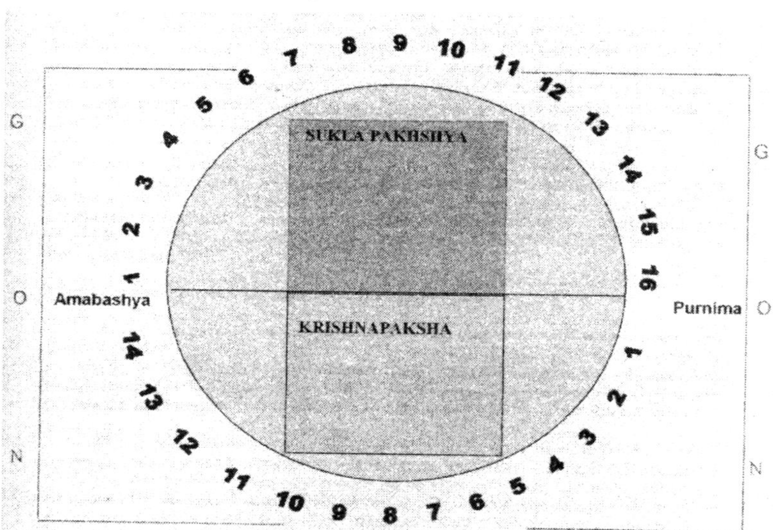

Concept of Lunar Calender

The inhabitants of Sundarban are very much dependent on the nearby surrounding, *i.e.*, natural resources for maintaining their livelihood. They catch various kinds of fishes and crabs in the creek and river which depends on the position of moon. According to their traditional experience the lunar position of the calendar indicates the position of river or creek water level during high tide and low tide. The quantity of fish or crabs decrease or increase according to the velocity of the water current and the tide and ebb which is based on the lunar position and is locally known as "*tithi*".

Each month is divided into two halves or fortnight (*paksha*). One fortnight (*paksha*) is brighter, *i.e., Suklapaksha* (full moon) and the other fortnight is darker, *i.e., Krishnapaksha* (no moon). The fortnight periods, the darker moon is known as (*amabshya*) and the full moon is known as (*Purnima*) respectively.

According to their concept each paksha is divided into three categories, as per their experience of *Suklapaksha*, the 1st-5th days are locally known as "*Gonmukh*". During the *Gonmukh*, the river and creek water levels are high and saturated; current is gradually decreased with high tide in the river or creek. It is the best time for catching of fish and crabs. Within this *gonmukh* period 14th day of *Krishnapaksha* to 5th day of *Suklapaksha* is considered as *Barogon*.

The second part is mentioned as "*Marani*" which indicates similarly from 6th-9th days of the Suklapaksha. At the time of *marani* or *marakotal*, the best time for *Bagda* (meen), collection is done by the female folk at the river side. During this period river and creek are generally calm and quiet and the jungle goers are back home.

The third part is locally known as "Barogon" or "Bharakotal". It is considered from the 14th days of Suklapaksha to 2nd day of Krishnapaksha. It is considered from the 14 days of Suklapaksha to 2nd day of Krishnapaksha, During this period the river or creek water level reaches at an optimum level and there is a strong under current. It is the best time for catching different types of fishes and crabs.

Similarly during the darker moon they observe the term locally known as *Krishnapaksha*. Again this period is divided into three parts each according to their experience. From *Krishnapaksha* 1st to 5th day, it is locally known as *Gonmukh*.

In this period the river or creek water level is high and the surface water becomes saturated. It is the best time for more collection of fish and crab. "*Marani*" or *marakotal* starts from 6th to 9th day of the both *paksha*. During this period the river or creek is generally calm and quiet. It is observed that in both the "Marani" they prefer to be back at home because the time is not suitable for the catching of fish and crabs.

The third part of the calendar is similarly known as "*Barogon*" or *Bharakotal*, which starts from the 10th 14th day of the Krishnapaksha, to 3rd day of Suklapaksha. It is the best time for catching of fish and crabs because of strong under current. The people follow the traditional way of cylindrical system, which is part and parcel of collection of aquatic resources like fish and crabs from river and creek.

Wind variation and resource availability

At the time of honey collection and other resources from nearby reserve forest, they mainly depend on wind variation as per their experience. They consider wind movement broadly into two type's *Dakshiner hawa* (southward wind) and *Utture hawa* (northward wind) According to their knowledge, in the Bengali calendar the month from *Chaitra* to *Aswin* (April-October), the period is considered as *Dakshiner hawa* and is the best time for honey and fish collection.

During rainy season, *Shravan* and *Bhadra* months are not suitable for any type of collection, because wind blows from east ward direction, locally known as *puper hawa* (eastward wind) The river turns very rough and water level becomes very high, with under current and whirlpools. The situation does not permit the voyage. During this time they generally prefer to go to *Khari* (creek) for collection. Similarly, they consider the month *Kartik* to *Falgun* for *Utture hawa*. During this period they collect different types of fish, crabs, etc. Water level at the river and creek is calm and quiet and is considered suitable time for collection. During the *Utture hawa*, the two months of *Agrahayan* and *Poush* are not suitable for collection because west ward wind occasionally blows in these two months, and is locally known as *Paschimear* hawa (westward wind). During this time river turns very rough and strong under current, and there is damage of road and embankment of river.

The knowledge of wind variation and collection of honey is done by the *Bouley* who acquires some knowledge on the day of Shree *Panchami* and *Kojagori Laxmipuja*. They check and test the direction of wind's blowing on the night which they consider as "*Prohor*". Time around 9 p.m. is considered as *Protham prohor*, midnight is considered as *Dittiya prohor*, and 3 a.m. is considered as *Tritiya prohor* accordingly. As per their experience, they can locate the beehive, based on wind blow on that particular night. Generally the people practice an age old traditional ritual in the night of *Sree Panchami*, when they put on a lighted earthen lamp in an earthen pot and follow the direction of wind. The movement of the flame indicates availability of beehives. The *Bouley* organizes this ritual but other members also follow this ritual either in the night of *Kojagori Laxmipuja* or during Sree Panchami. They consider that the bee always moves against the wind, *i.e.*, "leeward".

At the time of honey collection in the forest, people observe the swarming and flying of bees. As they know from their experience, flying bees always go against the wind. On the experience of flying bees they locate beehives in the forest.

The presence of tiger is also identified easily by the forest goers. In this case wind also helps them. The tiger also roams against the wind to hide its smell/presence. Thereby the jungle goer depends on blowing of wind.

Taboos and belief practices

The female members of the jungle goer and participating members also practise some beliefs and taboos mainly at the time of honey collection which are given below:-

Honey Collection-Honey collection is referred to as "*mahal*" and those who collect honey from jungle are referred to as "*mouley*". During the time of mahal the occurance of tiger attack on human life, *i.e.*, conflict between man and animal including number of injury case, missing case, etc., are found during the period. It is observed that beehives are chiefly found in trees like *Khalsi*, *Garjan*, *Bani*, *Geoa*, *Keora*, etc, and the best quality of honey is collected from *khalsi flower*. Those who collect fish are known as "*Jele*" and those who collect wood from jungle are called "*Bouley*". Collection of honey and wax is completely seasonal, and it starts from *Chaitra* (March-April) to *Vaisakh* (April to May).

The symbolic identity of *Banobibi*, *Dakhin* Rai, and *Shahajangali* plays a crucial role in the forest economic life of the people of Sundarban Biosphere. For a long time, the people are associated with the mother cult culture, and the local people appease the mother cult without any caste barrier. Before the journey and after the journey, the team members pay offerings and worship *Bonobibi* for her blessings to keep them safe from tigers. They believe that her supernatural powers can protect or save their lives from all sorts of hazards and unusual events. Hindus and Muslims worship the same deities despite belonging to different religions. The goddess of the forest "*Banobibi*" is worshipped in almost every village.

The female folk follow some taboos for safety of their family members at the time of *mahal* which are as follows:

They finish all daily ablutions like bathing, combing hair, applying of oil, use of soap, cleaning of house and cooking, etc. before sunrise.

Do not see any soot on the cooking utensils during the day.

Do not wear vermillion on their foreheads.

Do not squeeze wet clothes in the day and do not do sweeping of court yard and house, etc.

On Fridays, the women observe fasting and offer special prayers.

Do not do any form of stitching and washing

They cannot sleep during day time.

Those women whose husbands go for collection into the forest can have bath and apply vermillion only during darkness but before departure of their husbands and next only on their return.

Men also follow some taboos during *mahal* which are as follows.

They shave and get their hair cut by the village barber for tackling any emergency death of family members in their absence. They do not enter the jungle on Fridays; instead they stay on boat. No onward journey is performed on Fridays for mahal as a mark of respect for Banobibi. The concept of purity includes no defecation on the ground or mother earth in the jungle but on leaf or other substance. Before journey they cannot eat or carry bananas, bottle gourd, gourd and radish during mahal. They also can not eat anything burnt or roasted; instead it has to be boiled, fried or cooked otherwise. These beliefs and practices, which have come into existence over a period of time, are strictly observed.

Contact and conflict situation

The areas where conflict situation occurs, the people generally develop contact with the Biosphere Reserve specially those who collect or are frequent visitors to the restricted zone. As a consequence of this unauthorized entry there inevitably results conflict among man, animals and authorities.

The table related to conflict situation is assessed below-

Sajnekhali Range

Not Under Tiger Project	**Under Tiger Project**
Herobhanga (8), Thakuran (4), Chulkati (8), Dhupibhasani (8), Lothian, Ajmalari (12), Saptamukhi (2)	Jhilla (6), Arbesi (5), Khatuajhuri (3), Chadkhali (4), Chamta (8), Matla (4), Gona (4)- [1to3] , Bagmara (8), Haldibari, Haribhanga (3), Gosaba (4), Chota hardi (4)-[1to 3], Mayadwip (5), Netdhopani (3), Prikhali (7), Panchamukhani (5).
INDEX ■ NO FISHING IN **CORE AREA** OF FOREST. ■ NO FISHING IN **SANCTUARY AREA** OF FOREST. □ PERMITTED AREAS OF FOREST FOR FISHING	The boxes above display the parts of the forest in the Sajnekhali Range, clearly indicating the specified areas where fishing is permitted. Figures in () indicate the number compartments in the respective section; figures in [ ] indicate those sections of the forest compartments for which permissions are granted.

As per notifications of the Forest Department some areas have been declared for catching fish and collection of honey from where the people are allowed to collect fish and honey on the basis of terms and conditions. But the area where contact occurs is generally the restricted areas or zones as per the version of the Forest Department. As per age old experience, people say that "*bag jekhane machh sekhane*" which mean that more fish are available where the tiger generally stay in core area because the fisher men are not allowed to enter into this area. But the people enter the core or undisturbed area for more collection which inevitably results in conflict faced between man and animals and between man and authority. On the basis of official record and field observation it is seen that conflict occurs between man versus animals and man versus authority. It is also observed that during the month of March to May the number of casuality rate is comparatively high.

Before going for collection into the jungle, the forest authorities demarcate some areas as restricted areas where people are not allowed to enter for collection. But usually the people who are permit holders or even without permit prefer to enter the restricted area because that area is undisturbed and they collect more from that area. When the people cross the area the forest authorities punish them and different types of offence cases are lodged against them, such as-

(1) POR-Prosecution Offence Records (caught killing tiger, in possession of tiger skin, honey collection, etc.-arrested and forwarded to court of law).

(2) COR-Compoundable Offence Records (illegal wood cutting, fishing, honey collection in the protected areas. Material seized, Penalized compounds for repeaters).

(3) UDOR-Undetectable Offence Records (violation by unidentified persons, all materials seized.)

Table-4 : Offence Commited at Sajnekhali Range (Upto January 2005)

Sl. No	*Forest Offence*	*2001-02*	*2002-03*	*2003-04*	*2004-05*
1.	POR	5	6	3	3
2.	COR	148	252	303	345
3.	UDOR	64	20	40	29
4.	Persons Arrested	18	11	-	-
5.	Collection of honey Kg.	9758	8065	10422	9566
6.	Bee Wax	713.5	624.7	646.85	542.94
7.	Men involved	287	276	334	553
8.	Passes issued	47	43	50	82
9.	Tiger victim-death	2	2	4	Nil
10.	Tiger victim-injury	1	1	2	Nil
11.	Compensation realized		60,655.00		134622.00
12.	Qty of Timber seized m3	24.13	9.027		9
13.	No of Dingi seized	38	13		4
14.	No of mech. boats seized	4	Nil	Nil	Nil

The Table-4 regarding the collection of honey was collected from the Sajnekhali Forest Range from 2001 to January, 2005. From the Table it is clear the tiger victim death and injury cases occur between man and animal. Compensation collected, timber seized, Dingi seized, etc, by the forest authority reflect the conflict between man and authority.

Table-5 : Forest Protection (Annual Report (2003-2004) of the SUNDARBANS Tiger Reserve)

Sl. No.	FOREST OFFENCE	2001-02	2002-03	2003-04
1	POR (No.)	14	15	22
2	COR (No.)	384	640	679
3	UDOR (No.)	177	189	173
4	No. of Persons arrested	59	29	69
5	Incidences of firing by staff (No.)	3	12	36
6	No. of Offenders died	Nil	Nil	Nil
7	No. of Offenders injured	Nil	Nil	Nil
8	No. of Forest Staff killed	Nil	Nil	Nil
9	No. of Forest staff injured/assaulted	Nil	Nil	Nil
10	Total Compensation realized (Rs.)	233,776.00	188,130.00	338,816.00
11	Quantity of timber seized	49.253cu m.	55.6213 cu.m	26.134 cu.m
12	No. of Vehicles seized	Nil	Nil	Nil
13	No. of Dingis seized	145	120	127
14	No. of mecahnised boats seized	13	11	6

The above table highlights the different types of offence cases year wise which is taken from the Annual Report of the Sundarban Tiger Reserve (2003-2004) and from this it is clear that conflict occurs between man and forest authority.

It is observed that a number of inhabitants have lost their life and some of them have been injured by the tiger and crocodile. Here, the time period is considered categorically in three sets from 1990-1995, 1996-2001 and 2002 afterwards. Again the cases considered loss of life by the tiger and crocodile in different villages. Both male and female folk have been considered in the Table given below-

Table-6 : Tiger and Crocodile Victim Cases

Year	*Dayapur*		*Pakhirala*		*Jamespur*		*Lahiripur*		*Jelepara*	
	T	C	T	C	T	C	T	C	T	C
1990-1995	4	4	6	1	3	1	4	-	4	-
1996-2001	5 (1)	3	4	3	5	4 (1)	4	-	3	-
2002 +	3 (1)	2	2	2 (1)	2	1(1)	1	-	3	-
TOTAL	12	9	12	6	10	6	9	-	10	-

'T' signifies tiger; 'C' signifies crocodile. Figures in parenthesis depict injured persons

Some case studies and genealogical cases were taken from different villages which are related to the person or the family member's victims by the tiger and crocodiles. The members or the persons belong to the village like Dayapur, Pakhirala, Lahiripur, Jamespur and Jelepara where contact and conflict cases are more because the people are dependent on forest based resources. From the Table it is clear that in Dayapur village total 12 tiger victims and 9 crocodile victims have been taken. Out of 12 tiger victims, two are injured but still they are engaged in forest based economy. In Dayapur the 9 members, both male and female, lost their life from the attack of crocodiles. In the Phakhirala village 18 people were attacked by tiger and crocodiles. Out of the 18 numbers 12 people have lost their life by tiger and 6 people are attacked by the crocodile. Out of the 6 persons 5 have lost their life, one member is still alive. Similarly in Jamespur village total 16 persons are attacked by tiger and crocodile. Out of 16 persons 10 persons have lost their life by the tiger and 6 persons are attacked by crocodile, 2 numbers are still alive. In *Lahiripur* and *Jelepara* villages no crocodile attack cases are found but 9 and 10 tiger attacked cases are found.

All tiger victims are males except one from Jamespur village. At the time of *Bagda* meen collection in the evening that woman was attacked by the tiger at the river side. The tiger crossed the river and attacked her.

It is observed that crocodile victims, both male and female, are mainly at the time of Bagda meen collection. Females are

Sl. No.	Case reference with section of law	Name of Officer	Activities
1	Gosaba P. S.case no 61 dt 18.10.2k U/s 13/14 F ACT.		Aquatic illegal activities.
2.	Gosaba P.S C-22 dt.13.8.01 u/s 14 F ACT.	S.I P. Datta	Do
3.	Gosaba P.S.C-32 dt.2. 10.01 u/s 14 F ACT.	S. I A.K. Singha	Do
4.	Gosaba P.S. C-36 dt 21.10.01 u/s 14 F ACT.	S.I . B. N,Das	Do
5.	Gosaba P.S. C-41 dt 12.11.01 u/s 14 F ACT.	S.I A.K.Singha	Do
6.	Gosaba P.S. C-5 dt 14.2.01 u/s 394 GPC.	S.I B. N . Das	Robbery
7.	Gosaba P.S.C-21dt 12.7.02 u/s 364A/397 GPC	S.I A.K.Singha	Kidnapping of five persons murdered at Raymangal river.
8.	Gosaba P.S. C-30 dt .23.9.02 u/s 14 F Act	S.I R. K. Paul	Do
9.	Gosaba P.S. C-2 dt 15.1.03 u/s 14 F Act	S.I R Ojha	Aquatic illegal activities.
10.	Gosaba P.S. C-9 dt 26.2.04 u/s 395/397 GPC	S.I C. Banerjee	Robbery at Chamta
11.	Gosaba P.S. C-40 dt 24.9.04 u/s 13/14 F Act	S.I A. Majumder	Aquatic illegal activities.
12.	Gosaba P.S. C-45 dt 19.10.04 u/s 13/14 F Act	S.I C Banerjee	Aquatic illegal activities.

attacked by the crocodile mainly at the river side and males are attacked at the jungle area during the time of *Bagda* meen collection. From the above discussion it is clear that conflict occurs between man and animal (both tiger and crocodile).

During the time of field work some official records were taken from Gosaba Police Station in relation to aquatic illegal activities by the local and foreign poachers. (see page 142)

A very disturbing trend that seems to emerge is the recent pattern where entire families (including women and children) are taking to the jungle, all by themselves as a team, for collection of various items and they do not seem to mind an encounter or conflict with these wild animals. The kind of punishment meted to these kinds of offenders is quite soft-only seizure of nets, tools, tackles and collected items.

From the above discussion, it signifies that the human contact and the exploitation of forest resources continue to fulfill the human demands for occurrence of contact and conflict situation.

IMPACT OF TOURISM

One can not ignore the development of tourism in the area but at the cost of the loss of the ecological balance. It has been observed that, due to impact of tourism development the local people earn money in different fields like transportation, business, and for supply of the tourist needs like honey, crab, fish, etc. To fulfill the tourist needs like building, road, transportation, jetty ghat, watch tower, etc. provided by the authorities during the developmental work the local labourers are involved in the said work. At the same time the number of tourists come to this area and spends the mode of economy in the different fields for their satisfaction.

Besides, local people also satisfy the tourist's needs to provide the honey, fish and crabs which they collect from the same ecology, *i.e.*, Sundarban Biosphere Reserve. Besides, increase in the number of tourists and means of transportation caused environmental degradation like pollution, generation of garbage, soil erosion and dumping of rubbish, etc. The nature of exploitation of resources from the Biosphere by the human being in the name of 'Tourism' seems to be responsible for the rampant environmental degradation. From the above discussion it is reflected that the potential strength of the Biosphere come under the human demands.

The effect of human activity in developing infrastructure for tourism of Sundarban Biosphere is due to the increasing tourists which are posing a severe threat of pollution. There is no proper plan to dispose off the solid wastes that are being accumulated in the area, which finally finds its ways into the river or sea. The wastes contain chemical pollutants and bio-medical ingredients. Dumping of wastes from urban area, factories and boats continue to flow in the Vidyadhari River. It is difficult to assess the magnitude of such damage; especially in the biosphere and the inaccessible portions of the protected areas. The soil of the area is thin and poorly formed. Owing to frequency of country boat (*bhat-bhati*) and too much ferrying of launch services at the time of tourist season there is direct effect of soil erosion and the ecosystem is disturbed in the area by oil spillage, sound pollution, etc.

The significance of the Sundarban Biosphere plays a vital role between ecological aspects and biological aspects for our needs. The Biosphere saves the human habitation like city of Kolkata from different types of natural calamities. The mangrove vegetation protects the soil erosion at both the river banks and sea shore. The presence of mangrove vegetation is also suitable for fishes. From the ecological point of view, one can not ignore the importance of the Biosphere.

Conclusion

From the above discussion it is reflected that due to the impact of colonization over a period the area became over crowded and the inhabitants became a potential threat to the ecological condition. The area once again became a further threat to Biosphere Reserve for the influx of tourism. The local people are engaged to fulfill the tourist needs like arrangement of honey, prawn, crabs, different types of fishes, etc. which are collected from the Biosphere Reserve. On the other hand, to provide better facilities to the tourists, the authorities constructed hotels, roads, jetty, watch towers, etc. Besides, increase in the number of tourists and means of transportation caused environmental degradation like pollution, and generation of garbage, soil erosion and dumping of rubbish, etc. The people came into contact for biosphere resources time to time for their survival. The process of illegal collection turns into conflict with the authority and also with the animal due to entrance into the core area.

From the discussion it is reflected that the Biosphere Reserve is already under pressure by the immigrants from abroad and different corners of India. Moreover, the Department of Tourism is also taking steps for development of the biosphere as a tourist spot. Keeping in the view, in the name of 'development' of higher perspective, one finds both the negative and positive aspects as two sides of the same coin. We could not ignore the development of tourism but development should be in proper way and more emphasis on alternative economic scheme which will be a need based implementation. Tourism and sustainable development could be achieved only by the way of protecting the environment and conserving natural resources for the benefit of bio-diversity as well for our future generations.

References

Annual report, 2003-04, Field Director, Sundarban Tiger Reserve, Canning.

Bennet W. John, 1976, *The Ecological Transition. Cultural Anthropology and Human Adaptation.* Pergamon Press Inc. New York.

Chaudhuri A. B and D. D. Sarkar, 2004, *Golden Triangle of Eco-education Tourism*. Published by A.B. Chaudhauri.

De, Barun (Ed.), 1994, *West Bengal District Gazetteers. 24 Parganas*. Govt. of West Bengal, Calcutta.

Debnath H. S., 2004, *Sundarban Biosphere Reserve*. Botanical Survey of India, Kolkata.

Hunter, W. W., 1998, *A Statistical Account of 24 Parganas*. Govt. of West Bengal.

Mandol, Shasanko, 1995, *British Rajatye Sundarban*. Punascha.

Raha , R.K., 2004, *Indian Sundarban-An Overview*. Sundarban Biosphere Reserve. Wildlife Wing. Forest Department.

Ray Choudhury, B. K., 1980, *The Moon and the Net*. An. S. I., Kolkata.

Sundarban Bishayak Daptar, 2002, *Paschimbanga Sankhaya Souvenir*. "Banbibi Utsab". 2-8th January.

5

Delta Culture of Sunderban: A Case Study of Gosaba Block

S.S. Mishra

Introduction: An overview of Sunderban

Sunderban, the famous Delta of the Ganges is a land located in the heart of Bengal. The Southern, South-Eastern and South-Western portion of undivided 24 Parganas district is comprised of this region. However, the Sunderban as such is mainly the southern portion of the Delta. This includes altogether 4200 sq. kms. of reserved forest and 5400 sq.kms of inhabited areas covering 13 blocks of South 24 Parganas and 6 blocks of North 24 parganas districts.

This region is recognized as Sunderban Biosphere Reserve under MAB (Man and Biosphere Programme) to establish a formal mechanism for coordinating and integrating diverse activities of conservation, research and training for creating a better situation of harmony between man and environment. It has extremely rich diversity of aquatic and terrestrial flora and fauna. The flora includes highly rich halophytic vegetation, which is commonly known as mangroves. These are salt tolerant seed plants, which ranges from shrubs to tall plants. This Sunderban mangrove forest is the only *mangrove* forest in the world having the *Royal Bengal Tiger* as its indigenous population. Not only this rich and varied fauna of Sunderban Biosphere Reserve constitutes of 15 species of mammals, 8 species of birds and 17 species of reptiles out of these species some are rare and globally threatened animals found only in this region. Here special mention may be made of estuarine crocodile (*Crocodiles porosus*), Leopard cat (*Felis bengalensis*), common alter (*Ultra Intra*) etc. Hunter (1878) too has mentioned about prevalence of varied types of animals "Tiger, leopards, rhinoceros, wild buffaloes, wild hogs, wild cats, barasinga, spotted deer, hog deer, barking deer and monkeys are the principal varieties of wild animals found in Sunderban." The habitual degration and ecological changes in course of time has caused substantial changes in the faunal composition of the region as well. It is reported that some important animals like swamp deer, barking deer, javan rhino and wild buffalo etc, have become extinct. These ecological changes are attributed to various measures and experiments undertaken in the region.

Prior to some hundred years or so this region was uninhabited filled with dense forest and wild animals. "The sunrays even possibly could penetrate some of these forest areas. Though the old history refers that prior to the invasion of the Mughul in the Bengal, year 1128, these land of Sunderban areas were in the finest state of cultivation and the villages in general were well populated. The number of mosques and other places of worship still remaining fully demonstrate its former splendour and magnificence" (Hunter pp. CXVIII). But afterwards this region became subjected to invasion and destruction, which changed the entire structure of this place. Afterwards when the East India Company began to rule this state in the year 1765, a few British administrators felt the utility of clearing this forest and proposed scheme for it. In the year 1783 Mr. Tillman Henckle, the judge and Magistrate of Jessore and superintendent of Sunderbans, in a letter addressed to Honourable Warren Hastings, Governor General of India, schemed to reclaim and populate this tract of forest. In the year 1811-1814, Lt. W.E. Morrison did great work by surveying the Sunderbans. Afterwards Lt. Hodges prepared a map, which has been the standard map of Sunderbans ever since. It was followed by other developmental works in the region undertaken by many other British administrators, who facilitated reclamation and settlement in other areas within the Sunderbans. However, Hamilton's approach in this context was entirely different and had a concrete visionary outlook to develop the region quite distinctly; which is now known as Hamilton's Island.

Gosaba the Hamilton's Island

The Gosaba block *i.e.*, the Hamilton's Island constitutes of 14 *anchals* and altogether 51 *mouzas*.

Gosaba block at a glance

Total Area	-	285.85 sq.kms.
Population	-	2, 22,764 persons
Male	-	1, 13,826 (51.10%)
Female	-	1, 08,936 (48.90%)
Total Families	-	46,962 only
BPL	-	19,853 (42.27%)
APL	-	27,109 (57.73%)
Schedule Caste Population	-	1, 30,355 persons (58.52%)
Schedule Tribe	-	19,435 persons (8.72%)
Muslims	-	14,035 persons (6.30%)
Christians	-	6,015 persons (2.70%)

The area of the Gosaba block is altogether 285.85 sq.kms having population of 2,22,764 persons. This population constitute of 46,962 families. Out of which 42.27% falls under below poverty category while altogether 57.73% families are economically sound kept under above poverty line. Here the Schedule Caste people are in the majority, which count 58.52% of total population while the Schedule Tribes are just 8.72%. Likewise Muslim and Christians are comparatively lesser in number having 6.30% and 2.70% respectively. These populations together represent the multi ethnic nature of settlement in the region.

When the population and geographical nature of the region, in terms of its Grampanchyat and mouza, is taken into account, very characteristic feature emerges. Out of total 51 mouza of Gosaba block, one mouza of Zel No. 20 named char Amlamethi is an island without any settlement, while rest all are habited islands, where different communities live together quite amicably. In some mouzas settlement of some particular communities are reported to be more concentrated. The schedule caste populations are thickly concentrated in Gosaba, Pathankhali, and Bipradaspur having more than 65% population while in Radhanagar-Taranagar and Satjelia their percentage is a bit less but higher than any other population. Similarly schedule tribes are found in almost all *Grampanchayat* but are in sizeable number in Gosaba, Pakhirala, Radhanagar-Taranagar and Satjelia; while in other areas their population is quite less.

Different mouzas under this block are mostly surrounded by different rivers.

Here transport from one place to other is mainly based on small-mechanized boats, which ferry people to different ghats. These services are mainly available during daytime, which become irregular after sunset. There are altogether thirty number of ghats under *Gosaba Panchayat Samity*. Out of which some ghats are duly constructed with proper facilities of alighting but most of the other ghats are temporarily made, where movement of people become very difficult especially during ebb time.

Communication within island

Similarly communication within the Sunderban is another important aspect worth mention. Communication from one island to other is mainly through water ways. But inside one island communication is not small. Here van rickshaws play important role.

Communication is a big hindrance in the Sundarbans. Waterways are the main channel of communication from one island to another or from islands to the mainland or vice-versa. Inside the island, rickshaw van is the main mode of communication. In Gosaba Island covering Satjelia, Dayapur and Gosaba proper, there are more than 700 rickshaw vans operational for carrying passengers from one place to other. Gosaba, being a big island counts altogether 21 van stands for smooth movement of people from one place to other within the island.

In Gosaba proper altogether 500 vans are in service to cater to the needs of local passengers. The charge varies from Rs.2/-to Rs.7/-depending upon the distance. Hence average income counts Rs.60 to 100 per day.

The van pullers are working under a union called Gosaba thana Rickshaw van chalak Samaittee, which regulates their entire gamut of activities including distribution of ticket, formulation of rules, their welfare measures and other socio-cultural activities.

Problems

A van normally costs Rs. 4800/-to Rs. 5000/-. Apart from it the condition of road is not in order, which badly damage its parts like tire, bearing, etc. The wear and tear is quite high. Likewise Rs.501/-is required to be deposited for new van puller for purchasing the route, on which he can carry passengers for his earnings.

During winter season the van pullers earn good amount due to arrival of large number of tourists to watch the jungle of Sunderbans. In other season, passengers come out quite less in number. As a result one hardly gets chance of two turns.

Recently attempts were made to bring mechanized vehicles for smooth and early movement of people. The van union vehemently opposed this move and observed strike against it in 2005. The van pullers fear that once such vehicles enter in the Gosaba, it will adversely affect their income.

Presently rickshaw van is serving as the life line for carrying people, materials and other belongings. On the one hand it is cheaper, similarly giving bread to good number of van pullers. But situation is likely to change sooner or later with the growth and development of Gosaba in particular and other adjoining areas in general.

HAMILTON'S ISLAND (GOSABA)

Emergence of new culture

Sir Daniel Makhinnon Hamilton, the chief of Makhnin & Mechenji Company after retirement, instead of going back to his native country Scotland, decided to remain amidst the dense forest of Sunderbans to establish his long cherished dream to set up a community undivided by caste and religion. He had a utopia to offer and he thought he had the magic store to build the dreamland (Bandhopadhyay: 2003 pp XIV). His all consuming passion made him the legend in the area. It is why this area Gosaba (a land of his activities) is also termed as "Hamilton's Island" or Hamilton's Gosaba.

Sir Hamilton was a utilitarian in his approach, who believed that the sole interest of economics is not only economic growth, but also the development of human beings (Raha: 2004 pp.39). In the year 1903 Sir Daniel took a lease from the British Government, the forests of the three Islands in the Southern most part of the Sunderban for deforestation and colonization. The lease was granted under the Large Capitalist Rules for three block of lands *viz.* Gosaba Island (approximately 830 acres in area), Lot 143 (approximately 3200 acres) and Lot 149 (approximately 4850 acres in area). Afterwards in 1909 these grants were supplemented by the lease of another island *i.e.,* Lot 148 (adjacent block exceeding 14000 acres). In this way the Hamilton's estate is comprised of four islands of a total acreage of more than 22,000 acres. When this vast area was acquired, the foremost job before them was its reclamations.

Reclamation and settlement

Sir Daniel Hamilton stepped in Gosaba and started reclamation work is the utter wildness of the jungle. He had to import labour from distant areas. Here the tribals were being brought from Chotanagpur, Bihar (Now Jharkhand) and Mayurbhunj area of Orissa. He made all necessary arrangements for all their requirements including provision of drinking water. The reclamation of land, however, was an onerous task and could only be carried out with great caution. The nature of the soil was such that only a small portion of the land about 1000 to 1500 acres could be taken at a time. A belt of say 120 feet. of jungle around the area was cut and earth dykes were set up in the belt and later, one after another of the numerous creeks and canals were bounded. Then work could be done, only in the dry months. When the area was enclosed, the jungles were cut, but the trunks were left standing. The rain came and washed the land, but the sluice boxes were made for drainage of the water in the enclosed area into the rivers during the ebb tides. The reclaimed land did not grow much paddy but as the land got good washing for some years, it seldom failed to yield bumper crops.

When the first portion of the land was successfully reclaimed, Sir Daniel invited outside people to settle down here. Unfortunately his invitation was declined. It happened because the common people had been badly deceived by unscrupulous zamindar, their officers and by the money lenders. As a result the benevolent mission of Sir Daniel failed to attract people. Moreover, he had also no intimate knowledge of them, their customs, and their mode of living, their indebted ness and their social conduct. In 1910 the population of the estate was about 900, including coolies; though land had been reclaimed to the extent of 3000 acres. The reluctance of people towards setting down is greatly attributed to the age-old distrust of the people and lack of communication to give them correct information. However, gradually people realized the benevolent gesture of Sir Daniel. The population of the estate recorded healthy upward trend with settlement of about 15,000 persons in the 1941 census. Now the foremost task before him was to facilitate the settlers to develop by their own effort through the concept of cooperation among themselves.

Co-operative movement

Sir Daniel launched the concept of co-operative system in Sunderban almost simultaneously with the ongoing co-operative movement in India. He was so committed to the concept of co-operation that he had carried the message throughout his life that if co-operation fails, there will fail the best hope of rural India. His various co-operative activities proved quite effective in developing Gosaba as a model village.

After reclamations, when he offered free land to the people in his estate he had expected simply nothing but one desire with the settler's that; they could not bring all their petty little divisions and differences. Here there would be no Brahmans or untouchables, no Bengalis and no Oriyas. Everyone would have to live and work together. This announcement induced people with some initial hiccup leading to settlement of quite sizeable population. These people (the tenants) were granted nearly 17000 acres land out of the total area of 22,000 acres in the estate. In all these endeavors Sir Daniel strictly adhered to the principle that not more than 25 bighas of land, which an able bodied man can at best cultivate, be settled to one person and that land never gravitated into the hands of a few. He successfully maintained his life's mission not to enrich a few but to help many (Majumdar, 1942. pp 577). Afterwards when Sir Daniel evaluated his scheme, he realized that there were very little or no changes in the economic condition of the tenants. "Sir Daniel was greatly disturbed to see that moneylenders were failing his work. He launched a crusade against them, and though the debts of the agriculturist with the moneylenders were found to be much more than their actual assets here, he saw all his people freed from the mahajan's within a week and the pest was once for all banished from the estate. This bold step worked wanders. By the repayment of the loans in statements with a fixed percentage annually, the people found themselves free of debts within a full year (Ibid pp. 574).

But all these efforts had limited impact on the people. He then resorted to introduce his long cherished dream of co-operative movement and thereby develop a feeling of share responsibility. Initially little response was reported because of unlimited liability of co-operation and the attraction of easy finance with the estate. "Great persuasion was required to start the first co-operative village society with 15 members in 1915. Sir Daniel provided an initial capital of Rs.500/-for the society, which formed the nucleus of group of rural credit societies, which sprang up later on. The credit movement began to spread. Afterwards these were federated into a credit union of the central type under the name of the Gosaba Central Co-operative Bank. The total number of rural societies by the end of June 1933 was 19 with 682 members and the accumulated reserve found amounting to Rs.11, 830/-. Unlike other co-operative societies, the system of issuing loans and their recovery was peculiar to this society. Long-term loans were issued. Generally loan was available for about five times in a year during the months of May to September and the purpose of these loans chiefly related to cultivation, purchase of cattle, maintenance and other unavoidable expenses. Its recovery was affected from the crops and more in kind than in actual cash (Bandhopadhyay: 2003 p: 12). In course of time almost the whole of the estate had been covered with village co-operative societies. Apart from it three more societies were formed with limited liabilities for the benefit of the people settled in the estate; namely

(i) The Gosaba Co-operative society (1923)

(ii) The Rice Mill (1927)

(iii) The Employees Co-operative Society Ltd. (Credit)

Here the co-operative society was meant to supply the members of all co-operative societies with the daily necessities of life at a cheaper rate by means of joint purchase. Likewise a co-operative paddy sale society was established to eliminate the middlemen, who used to control the price of paddy in the locality as the tenants were unable to go individually to distant markets for higher prices. This society served the object of pooling the supplies of members and disposing of the same in the best market. Similarly the establishment of the Jamini Rice Mill in 1927 was an organ of the sale society. It was meant for ensuring finished products and their marketing with maximum amount of profit for the tenants. The mill was named after Shri J.N. Mitra, the then Registrar of co-operative societies, Bengal as a tribute to the services rendered by him to the movement. Apart from the provision of rural bank, Dharamgola (1929) & one rupees currency etc, introduced by Sir Daniel, he also gave pace to the co-operative movement.

The co-operative system was developed with great pain. Here a mechanism was evolved so that even a small village society has a role in its operation and management. The Board of Directors of the co-operative banks were entrusted to look after the village welfare as court of justice. As per the provision, the High Court was the Directorate of the Central Bank together with the directorates of any six village banks. This body used to settle disputes apart from being court of wards and controller of the Dharmagolas (A store made in every bank for emergencies). Due to all these mechanisms, not a single criminal or civil case was recorded in Gosaba for over 30 years and even the Government survey and settlement operators recorded not a single dispute.

The members of the co-operative societies used to get together once in a month in the *Samabay Baithak* to forward their recommendations to the central welfare committee. There was a panchayat in every village. Anything like fifty to seventy five persons used to participate in the annual election of the five members making up the panchayat in every village. The proceedings were marked by liveliness without friction.

The co-operative movement was at its peak during Sir Daniel's regime. However, after his death in the year 1939, this movement continued in the guidance of Lady Hamilton. But it faced severe set back with her demise. Gradually Gosaba began to loose its earlier charm. After independence Sir Daniel's trust was vested to the state leading to further deterioration. Due to all these factors the frequency and level of co-operative movement has come down quite drastically. Yet its impact cannot be denied on the people even till date. Health is one such factor where due emphasis was paid till the very days of reclamation in the estate.

Grow more paddy movement: A path of success

The Sunderbans are strictly a rice growing country and paddy forms their great staple. Here two crops are sown in a year namely the autumn (*aus*) and the winter (*aman*). But the cultivators mainly depend on the cold winter paddy. The intensity of autumn cultivation has risen in recent year with the improvement of irrigation facilities.

In the beginning when Hamilton occupied this area for deforestation and colonization, situation was quite adversed. "The reclaimed land did not grow much paddy, but as the land got good washing for some years dibbling and later to yield bumper crops. In the early years dibbling and later broadcasting of paddy was resorted to and when the tree trunks were taken off, the land allowed ploughing" (Majumdar: pp 573). Low yield was a matter of great concern. Realizing its gravity and to improve the quantum of yield Sir Hamilton established a central model farm in 1919 to experiment with vegetables, fruits and especially to research on paddy and demonstrate the results with branch model farms attached to schools in every village. As a result of work at Gosaba, a distract in variety of paddy well known as *Gosaba 23* in *Patnai* was evolved.

It bears the distinct characteristic of withstanding drought and disease. This achievement revolutionized the economic condition of the farmers of the estate; as the new strain yields an average of 10 md. per bigha, while previously the yield never exceeded 6 md. Good yield encouraged farmers to grow more paddies. Apart from this an impetus of the estate in the form of holding paddy yields competition for the farmer's honour them for their endeavour also proved fruitful. Annually the estate used to award 40 prizes of the total value of Rs.2000 to the best cultivator of Rupsal Gosaba 23 Patnai and Kumragore strains in the ten divisions of the estate. "The largest yield recorded in the competition is 17 md. 38 sr. of Gosaba 23 Patnai per Bigha. This shows what a good yield the land can achieve".

Alongwith the paddy the farmers have started producing cash crops at large scale also including Moung, khesari, oil seeds and chilly, which is a recent development.

No.	Types of production	Cultivable Area	Production (yield) (7.5 Bigha–1 Hectare)
1.	Moong Dal	400 to 500 Hectares	1.2 to 1.5 quintal per bigha
2.	Khesari	3500 to 4000 Hectares	1.2 quintal per bigha
3.	Sun flower	250-300 Hectares	1.5 to 2 tons per Hectare
4.	Til	100 Hectares	1.2 to 1.5 tons per Hectare
5.	Tisi (Lin seed)	20-30 Hectares	1.2 quintal per bigha
6.	Chilly	3500 to 4000 Hectares	1.2–1.5 tons per bigha

Source: Agriculture Office 2006.

Government is also trying to promote farmers through different programmes like crop diversification programme, NWDPRA (National Watershed Development Project) in rain fed area etc., for these cash crops. Even then paddy is still the main crop produced by them. Recently the farmers have started producing second paddy crop as well with the help of lift irrigation. In Gosaba, there had been provision of storage of surplus paddy since the early years of their settlement.

Present data on yield/types of crops etc.

With the growth of high paddy production, there was necessity to store it properly especially for the lean period. Normally it was observed that farmers unable to store, preferred to sell their produce for very meager amount. While in the lean season they had to purchase it by paying quite exorbitant amount. Here it was realized by the estate to develop storage house for the farmers. It will keep their paddy properly and protect them from undue exploitation from the Mahajans. It was called Dharamgola or Grain Gola. It was meant to help the farmers in the emergencies with little amount of interest in the form of kind (paddy).

Like other co-operative enterprises the Dharmgola came into existence. Then there were altogether 36 nos of such units in the entire estate. The farmers keeping their paddy had to pay 25% per quintal paddy as annual charge. The farmers used to take out their share whenever they desired whether to be used as seeds for cultivation or for consumption These Dharamgolas worked quite successfully during Hamilton's reign even afterwards for some time.

In the post independence period things changed rapidly, estate was vested, Hamilton was no more and there was none to maintain it in the co-operative environment. As a result these units began to collapse gradually and lost their existence in no time.

Afterwards in the year 1965 a gentleman (political activist) named Sri Balram Bera of village Bagbagan took interest to revive this institute. He along with some other fellows started it in his own village, which got registered afterwards in the year 1969 (Regd. No. 120). Since then this Dharamgola is functioning quite smoothly, which has privilege to be acclaimed as only such type of society not only in the Hamilton's estate but in the entire West Bengal state.

This Dharamgola constitutes of 293 members including

Schedule Castes	71
Scheduled Tribes	12
Females	21
Others	189

Other Works

This Dharamgola is running in profit. Presently a community hall is being constructed over the existing Dharmgola to provide accommodation and space for meetings, etc. Entire construction is done by the profit earned from the interest of Dharamgola. Not only that, some portion of earning (profit) is used in other welfare activities too. In the year 1989, when floods affected the local inhabitants, free paddy was distributed to the suffering people. In this way this lone representative of Hamilton's legacy is still a very effective institute serving the needs of the people of that area.

What is Dharamgola?

Dharamgola or food grain gola is a big room duly covered and well protected to check entry of rats etc. Here paddy is kept packed in bags with proper naming and numbering. In Bagbagan there are two such big rooms (reservoirs) with a capacity of 25 thousands quintals to keep paddy. This year due to construction overhead paddy has been converted into cash by selling out the entire paddy.

The Health System

Sir Daniel Hamilton himself was very much concerned for the welfare and good health of people settled in his estate. In the initial stage of settlement enough provisions were made to render effective medical treatment. The first dispensary was established in the year 1910. At that time the European doctors were appointed but the uncongenial environment and lack of other infrastructural facilities compelled them to leave the place. Then the Gosaba area was a cholera prone area, which caused much casuality too. After the European medical experts, one Indian doctor named Dr. Nishi Bhushan Chakraborty joined the Hamilton estate. At that time there was one charitable dispensary and a big bungalow for the doctor, made of Burma Wooden structure. Once very severe torrent rain damaged the construction, this has been replaced by the existing concrete structure. Then the doctor used to visit different areas of the estate to attend the patient by riding on the horses. Dr Gopinath Burman, who is serving the estate from last 50 years or so also, had done this in the beginning to attend patient in the villages. But now being very old he prefers to treat his patients in the dispensary at Gosaba, where almost everyday about two hundred patients arrive from different villages of Gosaba block.

After independence the medical facilities in the Gosaba estate have improved significantly. Apart from Sir Daniel Hamilton trust, many non Governmental agencies have taken keen interest to improve the medical amenities. Here special mention may be made of Tagore Society for Rural Development, which has launched comprehensive health projects since 1978. Here well furnished 25-bedded hospital, out patient door and other medical provisions have been made to serve the people. A team of doctors and nursing staffs are available here round the clock to attend the suffering people. In addition one mobile health care service through motor boat has also been undertaken by this organization with the financial aids of Government of India. Some other institutions have also been undertaken by this organization specially the Christian missionaries and other non-Government organizations are also rendering their medical services to the people. The Health & Fainally welfare department of Government of West Bengal has also made adequate arrangements in this regard. This includes block primary health center (BPHC), two primary health centers and other units, which are functioning under the administrative control of block medical officer. This shows that in course of time the health system of the area has improved, so is the case with hygiene and sanitation. The rate of literacy and awareness has greatly contributed towards developing consciousness among them.

Drinking water

The Gosaba estate, being surrounded with rivers had experienced problem of drinking water since the very days of its reclamation. Sir Daniel had then made provision of drinking water by installing filtration plant and transportation of water from adjoining areas. With the growth of population additional resources were developed to meet the requirement. Presently almost every family is having at least one pond within his house premises, which serves all purposes of bathing, cleaning utensils and washing except drinking. The tube wells are the main source of drinking water. In the Gosaba there are altogether thirteen tube wells (twelve of Government and one owned individually). The pumping station is at Sonagaon, where from it is supplied to different water posts. Similarly Rangabelia has altogether eighteen tube wells with one pumping station at Bagbagan. These tube wells serve the major requirement of drinking water.

Awareness

The water posts located at different places to supply safe drinking water is a place of great social importance as well. Since the amount of supply is less than the minimum requirement of the people, the womenfolk assemble near to it to make queue of pots before the schedule time of water supply. This spare time gives them good opportunity of discussion and interaction with one another. Not only this, different visual advertisements, wall paintings and writings on the wall of water posts also become subjects of their discussion. Here they exchange their views openly on different issues including Aids, polio and other advertised matters. This often leads them to think and act accordingly. This awareness also has positive impact towards health consciousness and proper sanitation to avoid spread of diseases.

Sanitation

Health concern of people has greatly contributed towards maintaining proper sanitation as well. Of late, special drive has been launched in this block to make the area free from health hazards caused due to lack of sanitation and Health hygiene. Here due emphasis has been paid to install scientific low cost sanitary latrine in all the houses. The Tagore Society of Rural Development has been assigned this task to complete the target by 2007. As per the Government statistics the position of "the total sanitary campaign" till 2004, 2005 is as follows (Table–3).

Table-3 : Sanitary Campaign Status

i.	Target	37,356	100%
ii.	Physical performance for the year 2004	2,983	(7.99%)
iii.	Cumulative performance for the year 2004	10,769	(28.83%)
iv.	Balance	26,587	(71.17%)
v.	School toilets (Target)	108	——
vi.	Achievement (Cumulative) since inception	01	(0.93%)

Apart from this, two boats have also been kept on contract to shift sanitary materials to its destination points. The sanitary latrine are made of different models depending upon its low cost of Rs.450/-(Rupees four hundred fifty only) to very high cost of Rs.10, 000/-(Rupees ten thousand only). Government campaigning for total sanitation is mainly towards the low cost latrine, where some subsidy is also provided to the beneficiary. As per provision in the low cost latrine the Government provides subsidy of Rs.200/-(Rupees two hundred only) while rest Rs.250/-(Rupees two hundred fifty only) is borne by the beneficiary. Recently the government has enhanced an amount of Rs.25/-(Rupees Twenty five only) in the subsidy amount due to price rise in the materials used in the making of such low cost latrine. The entire block was surveyed by a term to count the number of households and their economic status to be grouped under two broad categories *i.e.*, families who are within APL (*i.e.*, Above Poverty Line) and another economically poor families; who are grouped under BPL (*i.e.*, Below Poverty Line). The detail is mentioned in the Table No.4. (As per 2001 information's)

Table-4

Sl. No.	*Name of Gram Panchayat*	*BPL*	*APL*	*Total No. of Households*
1.	Sambhunagar	1408	1515	2923
2.	Pathankhali	1158	1589	2747
3.	Bipradaspur	1355	1935	3290
4.	Kachukhali	1088	1472	2560
5.	R. T. Nagar	2377	2057	4434
6.	Amtali	1409	1899	3308
7.	Kumirmari	825	2504	3329
8.	Chota Malkhali	1517	2182	3699
9.	Bali I	937	1819	2756
10.	Bali II	1489	1948	3437
11.	Gosaba	1971	1761	3732
12.	Rangabelia	1193	1789	2982
13.	Satjelia	1518	1961	3479
14.	Lahiripur	1608	2678	4286
	Total	19853 (42.27%)	27109 (57.73%)	46962

The sanitation campaign was started in the year 1993. Since then the low cost sanitary latrines have been installed in many houses. However, from last seven–eight years this campaign has gained momentum. Even then it seems a remote possibility to achieve the target of total sanitation by 2007 as is evident from the Table furnished by TSRD, Gosaba.

In this way till date 52.47% work has been done so far. This trend also indicates people's concern towards sanitation and hygiene. It has also facilitated them to maintain cleanliness not only inside their houses but in the outside areas too. This improvement can very well be observed in their mode of living and other related fields. Needless to say that in this drive the education and literacy campaign has played a very effective role.

Education

Sir Daniel Hamilton was very concerned to uplift the quality of life of the people through various measures including the spread of education. He believed that character is the foundation of any success. He was anxious to raise the quality of the human resources through every possible means including education (Bandyopadhyay: 2003 pp.14). He made provision to establish at least one school in every anchal. In this way fourteen schools were opened in the estate. Here one superintendent was deputed to supervise the overall activities of these schools. Not only that, Sir Daniel made provisions of special incentives for the teachers as well, who were inclined to serve in his estate. This included free accommodation and ten bighas of land for settlement. Due to these factors, the number of schools increased gradually, so was the growth of students. There were some provisions to develop interest of students towards education and to realize its value as well. The student of all the fourteen schools had to appear in the annual examination held centrally at Gosaba. Here the topper students of different classes among all the schools were given cash incentive and prizes for their academic achievements. The lady Hamilton, wife of Sir Daniel Hamilton herself used to give prizes to the talented students individually. This influenced the students to perform better. These traditions remain continued for a pretty long time.

In the beginning when the educational institutes were being developed, one European lady Miss White served here as education superintendent. The lady took very keen interest to develop the academic environment in the estate. "In the census of 1941, the estate recorded a population of 15,000 with literacy percentage of 25 excluding the areas, which were reclaimed in the thirties" (Bandyopadhyay: 2003 p.15). At that time there were 28 primary schools, 3 Middle English schools (M.E. Schools) and one RR Institute. Here the role of RR institute is worth mention, established by the inspiration of Sir Daniel Hamilton. He was of the opinion that plain education is in no way helpful to the people, so he planned to initiate vocational courses and training for the students to earn their bread. As a result RR Institute *i.e.,* Rural Reconstruction Institute was established in the year 1932 with this intention to train people in different field of vocations. Here both theoretical and practical knowledge are being provided for attaining perfection in different fields of vocation. This experiment yielded very positive results, which influenced the dignitaries of other areas also, who visited the estate to witness its impact. "In 1937 Sir John Anderson, the Governor of Bengal visited Gosaba to see the rural reconstruction activities".

The student of this institute are given "ILA" degree *i.e.,* Art of Independent livelihood in different trades namely Agriculture, dairying, weaving, etc. Many inhabitants got awarded with this degree and were employed in Sir Daniel's estate. A format of certificate issued to the students by the institute is mentioned below.

(a)	Primary Schools	-	151 nos.
(b)	Junior High Schools	-	08 nos.
(c)	High Schools	-	24 nos.
(d)	Higher Secondary Schools	-	08 nos.
(e)	College	-	01 No.

The awareness towards education and improvement of character through education influenced the literary growth to develop such environment with other such activities including publication of magazines etc.

Little Magazine

Bengal had been a seat of literary activities since long. Sunderbans, being a part of Bengal have also experienced such activities for pretty long time. As per an estimate about 50 odd numbers of little Magazines have been published from this Gosaba region. Unfortunately a very few could survive.

Sunderbans Samachar is believed as the oldest magazine, which was published from Mazidbari under the editorship of Mr. Ramanand Maiti. Afterwards many more such magazines came out.

Out of all only two little magazines namely *Lekhak* and *Lavnakt* are currently published, while rest all failed to maintain regular publication after publishing two to three issues. Lack of fund, limited circulation, little return on investment are some of the limitations faced by the editors to maintain the publications.

Education by all means had been a priority. When the present scenario is observed, Govt. has launched number of schemes including mid-day meal for developing interest in education.

The scheme of cooked mid-day meal is presently implemented in all the 151 primary schools, where altogether 25,271 students are being benefited. Apart from it this scheme is going on in 46 numbers of SSKS also, in which altogether 2459 students are getting education. Due to all these efforts, the literacy rate has registered sound growth. As per 2001 census, the literacy rate in Gosaba block is 44.08 percent. Out of this the rate is comparatively higher among the males *i.e.,* 55.07 percent while it is just 31.44 percent in case of females. The sample study done also indicates a similar trend. However, the rising trend of inclination towards educating their children and realization of its importance is positive sign of development, what Sir Daniel had dreamt for the people of his estate.

In this way Sir Daniel's special impetus towards co-operative movement, his tireless efforts to improve the character through education, better health and sanitation facilities yielded good result. Though Sir Daniel failed to implement many other schemes due to his untimely death but whatever he did in the estate has facilitated to develop harmonious inter-ethnic relations and a sense of oneness among all the people settled there. This feeling of oneness as an island settler has also influenced them to live together and fight together against all the natural odds like flood, wrath of sea and above all the fear of wild animals. Gosaba being surrounded by mangrove forest under the Sunderban biosphere reserve has been the abode of Royal Bengal Tiger and other ferocious animals. The people having no other options have to survive in this situation facing the fear of Tiger havoc. This situation facilitated the people to get themselves adapted to this new situation.

Adaptation

Sunderban is a deltaic forests region. In and around these forests a good number of people live in the different islands across numerous small and big stream and creeks in the adverse situation. This place was reputed for its unique flora and found biodiversity as well as the characteristic adaptability of the species. Like different species, the Sunderban people have also adapted to this new situation in spite of all odds.

Here most of the population are migrant community, who left their homeland long back and settled here. They have developed their root here itself in the new situation to be acclaimed as Sunderban people and have developed a distinct cultural identity as well.

The tribal and other population of plain areas have adjusted themselves in course of time to this alien aquatic environment. Multiple channels of rivers and the forest situation has greatly contributed in shapening their living. The rivers are no longer a barrier to them. It has proved to be a effective natural resource for them. Here apart from traditional fishe folk community almost all other community is resorted to this occupation for their livelihood. These other people are also termed as fishing community.

Fishing is carried out largely amongst the creeks of the Sunderban. Most common varieties of fish are *parse*, *tengra*, *chingri*, *bhola*, *bhangar*, *bhetki*, *kan-magur*, *kai*, *bagda*, *chingri* (Shrimps) etc. Nets of different shapes and sizes are used, which are fixed in the water course of rivers or creeks with the help of big drums, floating balls etc., such nets include *binti Jal*, *berjal*, *khepla jal* etc.

Fishing is also operated from the boat. In such fishing boats usually 8 to 10 persons perform entire activities. They get divided into two groups-one group stays in the boat and sets its moving holding one edge of the net, the other edge rests in the hands of the other group who moves encircling a considerable area by the net and reaches the bank. The boat encircles a considerable area of the waters and the net covers the area-such encircled area serves as the catching ground for the fishes. These nets are called *berjal*. In the jungle creeks these berjal (nets) are fixed at the time of high tide during the low tide (*bhata*) the water recedes and large number of fishes are caught in these nets.

Apart from this females of different community collect *minbagda* (Tiger prawn seeds) from nearby rivers but of late some of them are found venturing inside forest also along with the males. Similarly Sunderban forest proves to be another major resource of their income.

Likewise this unique Sundarban situation played an effective role in developing cultural cohesiveness among the settlers as well.

The common language is Bangla spoken by all including the tribal. The other dominant cultural traits like food habit, marriage, birth, death rituals are largely influenced by the Bangla culture. The dress pattern, use of vermillion, bangles and other related traits reflect prevalence of Bangalee culture. The fair and festivals, religious observances show lot of similarity. Above all the common cultural bindings of the Sunderban is the protector Goddess of Sunderban *i.e.*, Bonbibi, who is the prime deity of the region being worshipped by all.

Emergence of Bonbibi and the tiger cult

Sunderban represents a beautiful scenic combination of numerous tidal rivers and wide stretch of dense forests. The tidal rivers form network of rivers and water courses. "The great trunk channels enter the Sundarbans from the north and are connected by innumerable distributaries, which, after endless bifurcations and interfacings, unite into large estuaries falling into the Bay of Bengal "(Hunter, 1973, p.10). This river system serves as the principal means of communication as well.

The dense forest of Sunderban is known for its rich flora and fauna. Here more than thirty principal kinds of timber are found in abundance. These include Garan (Salacia prinoides), Pasur (Carapa. obovata), Sinj (Cynometra. bijuga), Sundri (Heritiera. Littoral's) Hental (Phoenix paludoso), Jhau (Tamarixgallico–var–indica), Dal Karamcha (Pongamia glabra), Balai (Hibiscus tiliaceus) etc. "Every year large numbers of people go to Sundarbans to collect wood. These trees supply the largest quantum of their collection. These trees serve many other purposes also" (Niyogi, 1987, p 95). Likewise among fauna apart from other wild animals, the Sunderban forms a type of citadel for *bagh* (Tiger). The Tiger of this region is commonly termed as Royal Bengal Tiger. "Among the mammals, Bengal Tiger perhaps evokes maximum fear, awe, respect and enigma all over the world. It is considered as a symbol of power, stamina and raw beauty ... In Bengal, the children grow up listening to the folklores based on tiger and reading spine chilling accounts of man eating tigers or heroic tales of tiger hunting. The Secular Goddess Bonbibi protects those, entering Sundarban forest for livelihood, from the wrath of tiger." (Raha, 2004, pp 30). This Goddess Bonbibi is worshipped by the people as mother, who protects his child all along in the forest. There is a very popular saying prevalent here, which is chanted by the Bonbibi worshipper. Sy. Montgomery, an American lady Scholar, while studying the man-eaters of Sunderbans had recorded this version from one *Gunin* (knowledgeable person) named Mr. Phoni Gayen of village Dayapur.

"O mother, thou who lives in the forest,
Than, the very incarnation of the forest,
I am the meanest son of yours
I am totally ignorant
Mother, do not leave
Mother, you kept me safe inside your womb
For ten months and ten days
Mother, replace me there again
O, Mother, pay heed to my words."

Gunin are the magic man, whose services are taken by the forest goers for safety.

Gunin (magic man)

The people of the Sunderbans while entering in the forest take one man with them, who is commonly known as *Bauley* or *Gunin*. The *Gunin* is a very important person who is a spirit charged person having ability to overpower the howling tiger, check its venture and also drive away the beast from the area, where the people are engaged in a sort of technical action like catching of fish in the creek, wood cutting or collection of honey.

Every forest goer party includes a Gunin among its member. This man stands surety for them by his performances, which is a symbolic communication in the forms of coded structure. The Gunin usually utters a series of abracadabra, which people call mantra. Through the effective application of this, the Gunin tries to regulate the venture of the tiger in the following way:-

Chalani: The tiger will be driven away from the respective place, where the people are at work, to another jungle.

Jalani: The body of the tiger will start burning, the beast will feel itching, he will be feeling uneasy to stay there.

Khilani: The tiger will not be able to open his mouth hole-his jaws will be seized with cramps.

Due to these applications the tiger will be inactive in venture, likewise the forest goers will enjoy safe return from the forest after fetching good amount of forest product.

Gunin occupies good position among the forest goer party due to his extraordinary quality. But there are some cases, where he failed to succeed. Likewise in a few cases they too became victim of tiger attack. As a result the people sometimes have said that these Gunins' way of behaving is but a hoax. Even then tradition survives—no Jungle goer party dare to enter the jungle without the *Gunin*. The *Gunin* may fail, but his failures never negate his significance among the people of Sunderbans.

Apart from *Gunin*, some other sacred items are also being taken by the forest goers to protect themselves from tiger and other wild animals. Here special mention may be made of *Tabij* or *Maduli* of pir sahib. It is being brought from other places believed to be very effective device. Likewise recently the *Rumal* (handkerchief) of Bangladesh is very popular. These *Rumals* are brought by any fellow member of Sunderbans from boarder areas of Bangladesh. A man named Pratidan Mandal and some other people used to provide these items to the forest goers.

In this way apart from *Gunin*, some sacred items like *maduli, rumal* etc., are very much prevalent among the people, who venture inside the jungle for good catch of natural resources as well as safe return to home against all odds.

Encounter with tiger is very common phenomenon here. Tiger is so well associated with each and every affairs of the Sundarban that no one can remain disinterested in him. In land tilting activities, in the collection of wood, honey and in fishing inside jungle, the tiger appears to be a big hurdle, which is difficult to overlook. The fear of tiger remains constant among the common rural folks, who earn their livelihood from the forest resources. This is harsh fact that the Sundarban people cannot remain away from their forest ventures as they have very limited options with regard to economic activities. "The regular jungle entrants are economically too hard pressed to afford technological arms to encounter the tiger. And, tiger, to these people is not a living beast only, which can be tackled by using technological arms. For them the tiger may be Dakhin Roy, the deity incarnate or it may be the carrier animal of Bonbibi—it is somewhat a supernatural entity. Naturally, technological action against tiger will be of no avail. With such belief system acting is in their psyche, they hardly take technological action to encounter tiger. The people are, therefore, inclined to take resort to expressive actions, which are manifested through symbolic activities, to counter the beast, which is a symbol of supernatural power "(Niyogi, 1996, p. 16)". Elaborate rituals are associated with Goddess Bonbibi, which are normally observed by the worshipper to get her blessings and protection from the ferocious Tiger. "Religious behavior both non-verbal and verbal manifest in expressive action, which is observed in honour of a deity or deities, who are supposed to be the controlling agent of this kingdom of varieties of flora and fauna and who can protect the men folk from being the victim of these ferocious carnivores. Such symbols of metaphysical abstract power are represented in an anthropomorphic icons namely Bonbibi, Harithakur, Dakhin Roy etc "(Niyogi, 1987. p 109)". This tiger cult deity is seen with reverence by almost every one living in the Sunderbans, irrespective of their distinct caste, creed and religious affiliation. This secular approach facilitates a binding force among them with commonness of distinct culture.

The fear of tiger havoc remains constant for the Sundarbans people, who earn their livelihood from the forest resources. The myth of Saviour Goddess Bonbibi and how she saved the unfortunate boy Dukhe from the wrath of Dakshin Ray, are told and retold by the people off and on to glorify her power and value. Who is equally revered by almost every man of the Sunderbans irrespective of their caste and creed? That is why a Hindu has no hesitation to bow down at the altar of Bonbibi who, from her name appears, has attachment with Muslim myth and likewise a Muslim forest goer has all the eagerness like his fellow Hindus. This syncretic feeling has added another dimension to the distinct cultural identity of Sunderbans.

Bonbibi has now become a household deity of Sunderban. Almost every community, respective of their caste and creed affiliation pay utmost honor to her; who is directly or indirectly related with forest affairs. She is worshipped in different places at different levels *i.e.,* individual level and village level. Likewise Bonbibi is worshipped even inside the forest by the different communities. At individual level she is revered either symbolically or by constructing some temple like structure depending upon one's economic conditions. The icon of Bonbibi are made by the concerned family either elaborately along with other icons or just shaped with soil. Similarly she is worshipped separately or along with other traditional deities. It so happened because Bonbibi is considered as an integral part of their tradition, who has been assimilated with them without any discrimination. She is worshipped regularly by the head member or the bride of the family with offerings and blowing of conch shell. Apart from it, village level annual celebration of Bonbibi is also prevalent. The ideal of Bonbibi along with her aids are constructed by raising contribution from each and every family of the village. The sacred food offered to Goddess Bonbibi is shared by members of different community. Similarly she is worshipped by forest goers inside the forest also. Apart from it the celebration of Bonbibi is observed in some other places also like Sajnekhali forest office (Gosaba), river side near the Dayapur Ghat, and some forest resorts. In all such places religions performance are observed by a knowledgeable person or any elder member of the community. Normally the services of Brahman priest are not required. However, in some cases they are reported to be invited. Sri Mohit Sanyal of Dayapur, a Brahman priest told that he had been invited by forest officials a few years ago to perform Bonbibi Puja. Sunderban people are deeply devoted towards Bonbibi for their well beings and existence. Apart from being Savior Goddess, Bonbibi is also worshipped as goddess of illness and even in other forms also depending upon one's dedication and divine instructions. Here some case studies have been incorporated to substantiate this fact.

Tribal Bonbibi Worshipper **Person-Mangal Sardar**
Case Study 1 **Dayapur Village**

Mangal sardar, a tribal of Bhumij Community aged about 70 years an old settler of Dayapur village under Gosaba block. His great grand fatherlate Golak Bhumij had come here along with a group of tribals from Bhila Basatpur village of Ranchi district, Chotanagpur (Bihar) now under Jharkhand state in connection with reclamation work of Sunderbans. At that time he was granted 25 bighas of land for livelihood. Presently Mangal Sardar possesses only 3 bighas of land. He was elected as head of the community to control and guide the community members according to the traditional norms of Bhumij society. He was very successful in this job as well. When his first wife died issueless, he married another tribal lady of Similpara village. Afterwards he brought another lady from Uludanga village as *Rakhani* (kept). The tradition of keeping *Rakhani* (kept) *i.e.,* a woman other than married wife by influential and rich persons had been an old practice in Chotanagpur. Being man of that region Mangal Sardar too did this, but it is quite rare in Sunderban.

Initially woodcutting was main occupation here but now they depend mainly on cultivation for their livelihood. Apart from it, they often enter in the forest located very near to their settlement, to collect wood, honey and catch fishes depending upon their requirements. Like any other forest goers, Mangal Sardar too used to worship Goddess Bonbibi for protection and safety from the wrath of Tiger.

Here there is a common practice that almost every forest goers family makes one *than* (small hut for Bonbibi) jointly or separately, where on one elevated platform icons of Goddess Bonbibi and her associates are kept for regular worship. In the locality of Mangal Sardar, two separate families; one of Munda tribe named Sobha Ram Munda and another of Bhumij tribe named Sagar Sardar have constructed separate *than* of Bonbibi. While three other families of Bhumij community namely Jotai Sardar, Panchu Sardar and Mongal Sardar together have constructed one *than* to worship Her jointly. They observe annual celebration of Bonbibi once in a year during *Magh Sri Panchami*, (Jan–Feb), when special offerings are made and elaborate rituals are observed. The daily worship includes cleaning of *than* with cow dung in the morning, offering of *tulsi* and sacred water in the noon and lighting of *pradeep* (lamp) in the evening.

Mongal Sardar is an ardent follower of Bonbibi, who normally performs daily rituals. Likewise whenever he intends to visit forest, he bows down before Goddess for the protection and safety of life. He believes that it is the Goddess Bonbibi, who has saved them till date. He opines that so long her blessings are with them, nothing adverse can happen in their lives.

Case Study 2 **Person–Haripad Das**

Pakhirala-Village

Bonbibi as goddess of illness

Bonbibi is believed as mother Goddess of *Atharo Bhatti Desh* (land of 18 tides), who is Savior Goddess, protects forest goers from the Tiger and other wild animals. Apart from this general notion, her other manifestations have also emerged. Here she is believed not only as a protector but a great heeler as well, who cures sufferers from different diseases, fulfills desires and solves various other problems of people.

One such *than* (temple) of Bonbibi is located near Pakhirala under Gosaba block. A few years ago Haripad Mandal, a migrant of Bangladesh has constructed it in his own house premises. Once he dreamt that Bonbibi is instructing him to construct one abode for her worshipping. Accordingly he did, where she had desired for it. Afterwards she demanded one pond for her bathing and a tree for her rest, which he completed in no time. In the beginning one Muslim fellow Md. Haffijuddin was being invited to perform rituals here including recitation of holy book "Bonbibi Zahurnama". In other special occasions also Md. Haffijuddin used to visit to perform her offerings. But now the wife of Haripad a Mandal, Srimati Subala Mandal performs all the rituals. She has devoted herself in the services of Bonbibi. Now she is called as *Dasi* (the slave of Bonbibi).

In this *than*, Bonbibi has been adorned in one small mud house with duly covered roof along with icons of other Gods and Goddesses made of soil. Here the icons kept on the left side represents Goddess *Manasa*, the right side icons are termed as *pesa pesi thakur*, and the icon placed in the middle is the central deity Bonbibi. In front of all these icons, necessary items of offering including Conch shell, *Ghanta, Sindur,* pots, flowers, incense etc., are kept for daily worship, to be performed by the *dasi* herself. Apart from it, numbers of plastic puppets are kept in one corner of the *than*, called as *cholan*. It is nothings but offerings made by the visitors, whose desire are fulfilled by the grace of Bonbibi. Just in front of the *than,* there is one open roofed sitting place with a *chulha* (oven) in one corner, normally used to prepare sacred food.

Bonbibi is worshipped here especially on Saturdays and Tuesdays, when visitors assemble in good number. Here no worshipping is performed on Sundays, Wednesdays and Friday. During these days even the door remains closed. Among the visitors the numbers of female worshippers are comparatively more than the males. On worshippers days the *dasi* gets invocated, when her voice and behaviour becomes different. It is believed that Goddess Bonbibi has influenced her body so whatever she says or even does are nothing but the act (instructions) of Goddess. During this state, visitors approach her one by one to place their problems. Here *dasi* gives them necessary instructions to be followed. Accordingly the visitor abides by the instructions and visit again with *cholan* (plastic puppet), when their desires are fulfilled. Apart from daily worship, special annual celebration is observed here on first day of *Magh* (Jan-Feb), when special offerings like *Suji sweet, payes, chini cholan*, etc are prepared.

One another important feature of this place quite uncommon is putting of *dhil* (small stone). It is believed that a person ties it with some *manta* (desire). When it is fulfilled, he removes it after paying necessary offering to Goddess. Many instances of fulfillment of desires, blessings with offspring, cures from different diseases etc., have been reported.

(1) Md. Arsad, a van puller of Pakhirala village once developed very serious headache. He took many medicines but had no relief. Then he visited this *than,* bowed before Goddess and soon he found himself fully cured. Later he came and made offerings to her.

(2) Mr. Satranjan Das of village Virajmoni married a girl of Gosaba but didn't get any issue even after five years. So he came here along with his wife to make prayer before Goddess. Afterwards he got blessed with a baby. Being pleased, he went to thank her and made special offerings like *payes, Khai* and fruits etc for her blessings.

Case Study 3

Bonbibi worshipper (Bonbibi worshipped together with tribal/Hindu deities)

Tribe–bhumij, nature of worship–family level

Type-idol worship, village-Katakhali (4 kms south of Gosaba)

Katakhali is a multi-ethnic village. Like any other areas of Gosaba block, here too Bonbibi worshipping is also very common.

There are many houses, in which Bonbibi is worshipped along with other traditional deities. One Bhumij man named Nitai Sardar aged 75 years old that their ancestors had come here for reclamation work. Then the entire area was infested with wild animals. Bonbibi, the Goddess of forest was their only hope. Since then she is being worshipped by them. Earlier an idol was not worshipped rather symbolic worship was common practice. But these days idols have became an essential part. In his house, Bonbibi is placed along with idols of some Hindu Gods and Goddesses namely *Sitala, Lakkhi* etc., Everyday his daughter in-law Smt. Kajal Sardar performs rituals with offering of *bhog* (sweet, fruits etc) and burning of *dhup*. The idol of Bonbibi prepared once is kept for five years. After that it is replaced with new one during annual celebration. At that time special rituals and offerings are made to be distributed among all invited friends and relatives.

In another Bhumij family a man named Makhan Sardar; in spite of his poor condition has constructed a separate *puja ghar* (worship house) for Bonbibi. Makhan Sardar and his wife Jangali Sardar are ardent followers. One day he dreamt that mother Bonbibi intends to arrive in his house. She told "*Tomar Ghare udbo*". It happened some 26 years ago. Soon both made up their mind to construct one thatched house for Bonbibi. Here symbolic icons of Bonbibi along with Hindu deities namely *Mahadeo,* Goddess *manasa,* a trident and their tribal deity called Baghrai (deity of forest) are kept together. These symbols are nothing but small rounded soil structures of about 6 to 8" inches height of almost equal girth are placed in series. These deities together along with Goddess Bonbibi are worshipped by them with great devotion and dedication.

Case Study 4

Bonbibi and Harisanyasi

(Deities of two cults worshipped together)

Place–Jele Para (Rajbanshi Para), Community–Jele or Rajbanshi (Hindu)

Harisanyasi is the deity of Jele (Rajbanshi) community, who migrated here from East Bengal (Bangladesh). Kachu Rai, a community member had brought some soil from the temple of Harisanyasi located there and built one small *than* of Harisanyasi, at Rajbanshi Para. Along with this deity, an icon of Bonbibi was also placed by them for daily worship. Afterwards some members objected that Bonbibi is a Muslim deity, which should not be placed together, so separate *than* side by side was made. Presently both structures are destroyed and vacant land is lying, where the members worship her symbolically.

Jele being fishing community often visit the jungle. The cases of tiger killing are quite high among them. Even in Jelepara, many widows are found who lost their husbands in the Jungle. The practice of keeping *Mangal Ghat (sacred pot)* in the name of Bonbibi is prevalent among them. A *Mangal Ghat (sacred pot)* is kept, whenever any members of the family leave for forest. It is done for the safety, security and safe return of members by the grace of Bonbibi. Here *ghat (sacred pot) is* worshipped by performing the same rituals *i.e.,* burning *dhup* and observing some restrictions. Likewise in the evening *jagaran* is observed. After the safe return of members Goddess Bonbibi is worshipped with elaborate rituals and offerings.

Shitla and Manasa *Jagaran* is a form of religious singing; where womenfolk assemble near the *than* of Bonbibi and sing the songs in her name. It is worth to mention that along with Bonbibi, other Gods and Goddeses especially goddess Manasa, Goddess Sitla, etc, are also remembered by them in their songs. This assimilation is quite unique, which indicate acceptance of Bonbibi as secular deity without any discrimination.

Other Cultural Traits

Apart from all these cultural characteristics; some other such features symbolizing Sunderban culture is found reflecting in their marriage system also.

There is reference of marriages within own community but intercaste marriages are also very common here. Here another important point worth mention is marriage distance. Here a striking character is found that Sunderban people have great affinity towards making alliance within the region itself.

Marriage Distance

Sl.No		*Total No. of marriages*	*Within Gosaba block*	*Beyond Gosaba but in Sunderbans*	*Outside Sunderbans*
FEMALES					
1.	Pondro	18	17	-	01
2.	Mahisya	10	06	03	01
3.	Tribal	09	06	02	01
4.	Brahman	02	02	-	-
5.	Muslim	07	02	03	02
6.	Jele	04	01	01	02
	TOTAL	50	34(68%)	09(18%)	07(14%)
MALES					
7.	Pondro	23	21	02	Nil
8.	Jele	11	08	03	Nil
9.	Mahaisya	10	09	01	Nil
10.	Muslim	06	00	05	01
	TOTAL	50	38(76%)	11(22%)	01(2%)
	GRAND TOTAL	100	72(72%)	20(20%)	08(8%)

Marriage distance is an important aspect of cultural linkages. Sunderbans being a distinct Deltaic region has experienced unique situation, as entire population has come from different places to this land. The long association of people here and dissociation from their ancestral home has facilitated to develop a great feeling of cohesiveness among themselves. When we analyse the marriage distance some distinct impression emerged.

The Sunderban people irrespective of their caste, creed affiliation considers himself to be a part of it with commonness among themselves. To them Sunderbans is a distinct world different from other places due to its unique climate, ecology and cultural environment. There is general tendency among almost all the communities to make alliance in a nearby area. Problem of communication difficult lifestyle, rivers all around and other problems also play as important role in developing preference of marriage within their area. This trend is still prevalent; though among females it is not so. The data on marriage distance indicate that girls of this area are being taken outside the Sunderbans also but it is almost missing in case of males.

In the sample 100 cases were collected to assess the marriage distance. Among females Pondro, having major population in the region prefer marriages within the area *i.e.,* majority of marriages (17 out of 18) were held in different villages within Gosaba block. Likewise among Mahaisya though majority of marriages took place within the block but cases of marriages outside block as well as well as outside Sunderbans were also celebrated. The other communities like tribals Muslims, Jele also show similar trend. But among Brahman all marriages were held locally. Their less population and limited settlement are also important factors. In Ghosal family of Dayapur, it was found that all three brothers married three sisters of a family.

The marriage distance among males also shows some interesting features. Among them there is preference for local marriage. Here some of them marry in other communities as well. Among four different communities studied. Marriage within block area is quite high among Pondro, Jele and Mahaisya but it is not so among Muslim. Out of 6 marriages none were done within block. All took place outside Gosaba with one beyond Sunderban area. But interesting feature of this marriage is that out of six-four boys have left their own place and settled near their in-laws house in the Gosaba.

Conclusion

In this way it is evident that since the days of reclamation, Gosaba has witnessed remarkable changes and development in different walks of life. In the beginning when Sir Daniel Hamilton took over the charge in the year 1903, the area was uninhabited being full of forest. But he had a model concept is his mind of developing distinct culture through his endeavour and people's participation. After the reclamation, people of different walks of life were invited to settle down here. They were granted free land and other facilities. Along with all these, Sir, Daniel Hamilton introduced many other plans and programmes like the Gosaba co-operative society Ltd, the Dharamgola and above all introduction of one rupee note in his estate to develop infrastructure and provide people with other basic facilities. Apart from it due attention was paid towards providing better academic environment as well by opening basic and vocational institutes. Similarly provisions were made to render good health care to people through charitable dispensaries and free medical aids. These efforts had begun to yield positive results but before it could gain momentum Sir Daniel died in 1939 leaving the entire things in mid-way. After Hamilton, Lady Hamilton took over the charge to continue it but she too died soon. After her departure other family members failed to continue it for long. Moreover, with the independence of the country and vesting of estate the whole system faced severe jolt. It began to loose its value due to lack of proper attention, devotion and effective mechanism, which Sir Daniel Hamilton had dreamt for this land and the people.

Gosaba being part of Sunderban Biosphere Reserve is surrounded with forest. Here the limited option of economic activities compels people to substantiate their economy with wood, honey and fish available there. Here the wrath of tiger happens to be a great hindrance. However, the existence of Bonbibi, the goddess of forest proves to be of great help in protecting the life and property of people. The greater dependence on forest increased the intensity of Bonbibi worship. Initially symbolic worship was prevalent here but gradually it was being replaced by erecting *than* and making idols at family and village level. Any one directly or indirectly connected with forest prefers to have Bonbibi *than* in one's premises, to get her blessings and protection. Off late, other manifestations of Bonbibi have also come up, where she is revered not only as protector from tiger but also who protects people from other socio-economic problems including blessings with offspring's, cure from diseases and alike. Due to all these rare qualities, Bonbibi has become accepted as household deity, who is being worshipped along with other traditional deities. Hence Sir Daniel Hamilton's dream of model society, which remained unfulfilled, found some place in the present scenario. Here the secular goddess Bonbibi played very important role, who is being worshipped by all irrespective of their caste and creed. This has also facilitated to develop distinct culture.

References

Bandyopadhyay, A., and Anup Matilal, 2003, *The Philosopher's Stone*, Speeches and writings of Sir Daniel Hamilton, pub by Sir DM Estate trust Gosaba24 Parganas(S).

Das, A.K., and M.K. Raha, 1963, *The Oraon's of Sunderban*, special series No. 3 Bulletin of the Cultural Research Institute, Govt. of W.B. Calcutta.

Majumdar, S. B., 1942, *Estate farming in India*, pub. *In Indian Farming*, Vol. III, No.11, A.C. Dutta, Sir DM. Hamilton Estate, Gosaba, 24 Pgs.

Mishra, S.S., 2005, Ganga Sagar, Ek bar nahi bar-bar (Hindi) Pub. In *Dristipat*, year-1, issue-1, June,Ranchi (Jharkhand).

Mishra, S.S., 2005, "Pahchan Ka Sankat Jhelti Sundarban me base Jharkhand ki Janjatia" pub in *Dristipat'* July, year, 1 issue. 2 Ranchi, (Jharkhand)

Mishra, S.S., 2005, Sunderban Jivmandal Arakchit Kshetra"Bangal tiger (Bagh)ki Saranasthali" (Hindi) pub in *Dristipat*, August, year-1, issue-3. Ranchi (Jharkhand)

Montgomery, Sy, 1995, *Spell of the Tiger the man eaters of Sunderbans*, Houghton Miffin Company Boston, New York.

Negi, B. S., 1985, *Regional Geography of India* Kedarnath Ramnath, Delhi.

Niyogi, T.K., 1987, *Aspects of folk cults in South Bengal*, Memoir No.74 Anthropological Survey of India Calcutta.

Niyogi, T.K., 1996, *Tiger Cult of the Sundarbans*, Anthropological Survey of India, Department of Culture, Calcutta.

Raha, A. K. & Bikash Kantisaha, 2004, *A wonder that is Sundarban*, Computronics, Kolkata (W.B)

Singh, T. V., 1991, *Tourism Environment* (Nature Culture Economy) & Inter-India Publications, New Delhi.

6

Economic Condition of a Delta Village: A Study in Biosphere Reserve in Sundarban

Krishna Mandal and Subrata Ray

Economic system is defined as an integral part of cultural system. Economy is concerned with the production, consumption and distribution of different things. People utilize surrounding nature for their survival.

Background

The study is mainly focused on economic condition of the people of a delta village *i.e.,* Pakhirala village of Gosaba Block in West Bengal, in the district of South 24 Parganas situated in southern part of Sundarbans.

Surdarban is the largest delta in the world which is situated in the estuary region of the Ganges and Brahamputra river. The Sundarban area of India is situated in North and South 24 parganas district of West Bangal. In this area rivers are linked up by rivulets, creeks, etc. Different types of mangrove trees, animals and birds are found which represent the attraction of Sundarban.

The Pakhirala village is located in transition zone of Biosphere Reserve, as well as a tourist center and it has a sizeable population with different caste and religious groups.

The village is situated near Gomor River in the Gosaba block of South 24 Parganas district in West Bengal. The village consists of four hamlets (locally called as '*para*'), named as—*Purbapara, Paschimpara, Uttarpara, Dakshinpara*. The total area of the Pakhirala village is 3.5 sq. kms. The name of the Gram Panchayat is Rangabelia Gram Panchayat. The villagers used van rickshaw, cycle and boat as their conveyance to move from one place to another. The main road of the village is metallic, other roads are earthen and brick made. A Zilla Parishad Bunglaw is situated in the village. There are eight hotels/lodges. The office of the Forest Range Officer is also situated in the village. Main source of drinking water of this village is through pipe water and a tube well .Generally, most of the families have a sweet water pond which also serves as source of water. There is no electricity but two solar street lights and 49 solar domestic lights are found in this village. There are altogether 800 families in the village. The total population of Pakhirala village is 3681 of which male is 1896 (51.51 per cent) and female is 1785 (48.49 per cent). Among the total population, literate individuals are 2614 (71.01 per cent), out of which male is 1508 (40.96 per cent) and female is 1106 (30.04 per cent). The Pakhirala is a multiethnic village. The population of the village belongs to different religious categories, such as—Hindu, Muslim and Christian. There are many castes in Hindu religion, such as—Brahman, Khandait, Tanti/Tantubai, Mahisya, Subarnabanik, Napit, Kamar, Satchashi, Poundra, Namasudra, Bagdi, Rajbanshi, Turi. Among the Hindus—Poundra, Namasudra, Bagdi, Rajbanshi and Turi belong to Scheduled Caste. The Munda is the only Scheduled Tribe inhabiting the village. The highest 62.37 per cent families are found in Poundra Community.

Economic aspect

From various sources like De, (1994), Hunter, (1998), mentioned that before Independence different dynasties like-

Mauryan, Gupta, Pal, Sen, Mughal have ruled in the Sundarban. Lastly the Britishers came here and build up colony for economically and socially backward people. In 1903, Sir Daniel Mackinnon Hamilton took the lands of this region and reclaimed the area by clearing forests. After reclamation the lands were converted into cultivable land. He set-up granary, co-operative bank, poultry farming, health centre, school, agricultural and weaving training centre for benefit of the people. For clearing the forests and construction of river embankment labourers were brought mainly from neighbouring districts like Medinipur and the then Bihar (Jharkhand). Some immigrants also came from North and South 24 parganas district , Orissa and adjacent districts (Jessore, Khulna) of eastern part of undivided Bengal now Bagladesh. Those immigrants stayed in this area and engaged in agricultural activities. Simultaneously they also depend on forest and river based resources and engaged in different economic activities like–fishing, honey collection, wood collection, etc.

The present study shows the economic activities of the people of Pakhirala Village. It is reported that before migrating in the village the communities had some traditional occupation, The traditional occupation of Mahisya, Poundra, Namasudra, Rajbanshi, Satchashi and Khandait, is agriculture. Traditional occupation of Tanti/Tantubai is weaving, Brahman as priesthood, Bagdi as fishing and cultivation, Turi as basketry, Napit as cutting of hair, beared, etc. But at present different type of economic activities like-agricultural/daily labour, hotel business, petty business/shop, boat man, tourist guide, van rickshaw puller, tailor, quack doctor, tutor, servant, service, "*dhadra*" (embroidery) work, batik work, etc, are found along with/without their traditional occupation.

Here, it is essential to know the position of traditional occupation of the study village.

Table-1 : Situation of Traditional Occupation

Caste Name	*Total*	*Followed*		*Followed with other occupation*		*Not Followed*	
		No.	*per cent*	*No.*	*per cent*	*No.*	*per cent*
Brahman	09	1	0.06	1	0.06	7	0.45
Mahisya	99	30	1.93	39	2.51	30	1.93
Khandait	08	4	0.26	1	0.06	3	0.19
Tanti/Tantubai	281	-	-	-	-	281	18.06
Subarnabanik	02	1	0.06	-	-	1	0.06
Napit	17	2	0.13	2	0.13	13	0.84
Kamar	01	-	-	-	-	1	0.06
Satchashi	08	1	0.06	3	0.19	4	0.26
Poundra	1026	285	18.32	350	22.50	391	25.13
Namasudra	14	1	0.06	1	0.06	12	0.77
Bagdi	82	68	4.38	4	0.26	10	0.64
Rajbanshi	05	1	0.06	1	0.06	3	0.19
Turi	04	-	-	-	-	4	0.26
Total	1556	394	25.33	402	25.83	760	48.84

From Table 1, it is found that out of 1556 individuals of both sexes of different castes the highest percentage *i.e.*, 48.84 per cent are not following their traditional occupation. Next, 25.83 per cent are following their traditional occupation with other occupation and 25.33 per cent are following their traditional occupation.

Now, community-wise traditional occupation is shown in Table 2.

Table-2 : Community Wise Traditional Occupation

Caste Name	*Total*	*Followed*		*Followed with other occupation*		*Not Followed*		*Total*
		No.	*per cent*	*No.*	*per cent*	*No.*	*per cent*	*per cent*
Brahman	09	1	11.11	1	11.11	7	77.77	99.99
Mahisya	99	30	30.30	39	39.39	30	30.30	99.99
Khandait	08	4	50	1	12.5	3	37.5	100
Tanti/Tantubai	281	-	-	-	-	281	100	100
Subarnabanik	02	1	50	-	-	1	50	100
Napit	17	2	11.76	2	11.76	13	76.47	99.99
Kamar	01	-	-	-	-	1	100	100
Satchashi	08	1	12.5	3	37.5	4	50	100
Poundra	1026	285	27.77	350	34.11	391	38.10	99.98
Namasudra	14	1	7.14	1	7.14	12	85.71	99.99
Bagdi	82	68	82.92	4	4.87	10	12.19	99.98
Rajbanshi	05	1	20	1	20	3	60	100
Turi	04	-	-	-	-	4	100	100

Table No.-2 shows that all individuals (100 per cent) of Tanti/ Tantubai, Kamar and Turi are not following their traditional occupation. Next, 85.71 per cent of Namasudra, 77.77 per cent of Brahman, 76.47 per cent of Napit, 60 per cent of Rajbanshi, 50 per cent of Satchashi, 50 per cent of Subarnabanik, 38.10 per cent of Poundra, 37.50 per cent of Khandait, 30.30 per cent of Mahisya and 12.19 per cent of Bagdi are also not following their traditional occupation.

Regarding followers of the traditional occupation, the highest percentage *i.e.,* 82.92 per cent of Bagdi are following their traditional occupation. Next, .50 per cent of Khandait, 50 per cent of Subarnabanik, 30.30 per cent of Mahisya, 27.77 per cent of Poundra, 20 per cent of Rajbanshi, 12.5 per cent of Satchashi, 11.76 per cent of Napit, 11.11 per cent Brahman and 7.14 per cent of Namasudra are also following their traditional occupation.

Regarding followers of the traditional occupation, with other occupation the highest Percentage *i.e.,* 39.39 per cent of Mahisya followed this type. Next, 37.5 per cent of Satchashi, 34.11 per cent of Poundra, 20 per cent of Rajbanshi, 12.5 per cent of Khandait, 11.76 per cent of Napit, 11.11 per cent Brahman, 7.14 per cent of Namasudra and 4.87 per cent of Bagdi following their traditional occupation with other occupation.

In the study village 1208 individuals of both sexes are engaged in different types of primary occupation. The highest percentages *i.e.,* 45.36 per cent are engaged in agriculture/horticulture. Next, 8.02 per cent are engaged as van rickshaw puller, 7.94 per cent work as agricultural labour, 7.69 per cent are engaged as daily labour, 6.53 per cent are engaged in fishing/catching of prawn-seed/crab, 5.21 per cent are engaged in petty business/shop. Between 1per cent to 5 per cent are engaged in different economic activities, such as services (4.55 per cent), other private jobs, tutor, servant/nurse-maid. Below 1 per cent are engaged in following economic activities like: skilled labour, honey collection, carpenter, boat-man, batik work, private job in hotel, quack doctor, hotel business, other business, tailor, *dhadra* (embroidery) work, tourist guide, barber, making and selling of fishing net/trap, gold smith and priest.

Here, the economic activities like–hotel business, tourist guide, boat–man, private job in hotel are directly related with tourism and some other economic activities like–petty business/ shop, van–rickshaw puller, batik work, *dhadra* (embroidery) work are indirectly related with tourism.

It is found that majority of the people are engaged in agriculture/horticulture. Therefore, it is essential to know agricultural landholding capacity of the families of study village. The Table 3 shows the distribution of agricultural land.

Table-3 : Land Holding (In Bigha*)

No.of Family (total 800)	*Landless*	*0.1-2*	*2.1-5*	*5.1-10*	*10.1-15*	*15.1 and above*
	273	277	181	52	12	5
per cent	34.13	34.63	22.62	6.5	1.5	0.62

* *Bigha* = measure of land, equal to twenty *kathas* (one *katha* = 720 sq. ft.)

From this Table it is observed that highest percentage *i.e.*, 34.63 per cent families have agricultural land within 0.1-2 bighas. Next, 22.62 per cent families have agricultural land within 2.1–5 bighas and 34.13 per cent families are landless.

In this village people cultivate paddy crop in rainy season. Some of them cultivated double crops in a land annually where irrigation facilities are available through canals. As a double crop they cultivate other varieties of paddy. They also produce 'moong dal'(one type of pulse), 'khesari dal'(one type of pulse), linseed, sunflower, chilli, potato, etc. Few of them cultivate sugarcane, watermelon, etc.

In the study village 815 individuals of both sexes are engaged in different types of secondary occupation. The highest percentages *i.e.,* 40.49 per cent are engaged in fishing/catching of prawn seed/ crab. Next, 25.88 per cent are engaged in agriculture/horticulture, 15.09 per cent work as daily labour, 7.23 per cent are engaged as agricultural labour, 4.90 per cent are engaged in petty business/ shop, 1.34 per cent are engaged honey collection. Below 1 per cent are engaged in following economic activities like van-rickshaw puller, skilled labour, tutor, carpenter, making and selling of fishing net/trap, maidservant/nurse-maid, tailor, private job, priest, tourist guide, boat-man, batik work, quack doctor.

Here, the economic activities like–tourist guide , boat man are directly related with tourism and some economic activities like–petty business/shop, van rickshaw puller, batik work are indirectly related with tourism.

It was also found that there are some economically poor families who have to engaged their children (0-14 years) and aged people (60 years and above) for their livelihood.

Now, the economic activities which are directly related with tourism are showing in Table 4

Table 4 : Economic activities directly related with tourism

Category	*Number of Persons*	*Per cent*
Boat–man	10	40
Private job in hotel	6	24
Hotel business	5	20
Tourist guide	4	16
TOTAL	25	100

Table 4 shows that 25 persons of both sexes are engaged in directly tourism related economic activities as their occupation. Out of 25 persons, the highest percentage *i.e.,* 40 per cent are engaged as boatman. Next 24 per cent work in private job in hotel, 20 per cent are engaged in hotel business and 16 per cent are engaged in tourist guide.

Here, the economic activities which are indirectly related with tourism are shown Table 5

Table 5 : Economic activities indirectly related with tourism

Economic Activities	*Number of Persons*	*Per cent*
Van–rickshaw puller	105	47.73
Petty business/shop	103	46.82
Batik-work	8	3.63
Dhadra (Embroidery) work	4	1.82
TOTAL	220	100.00

Table 5 shows that 220 persons of both sexes are engaged in indirectly tourism related economic activities as their occupation. Out of 220 persons, the highest percentage *i.e.,* 47.73 per cent worked as van rickshaw puller. Next, 46.82 per cent are engaged in petty business/shop, 3.63 per cent are engaged in batik–work and 1.82 per cent are engaged in *dhadra* (Embroidery) work.

Except that 16 part time workers of both sexes are engaged in different hotels/lodges for sweeping, cleaning, cooking, and as water bearers. Part time workers of other villages are also engaged in hotels/lodges.

It is notable fact that during peek tourist season of this area when the tourist does not find space in hotels/lodges, then they stay in the private houses near to the hotels/lodges or market area according to their convenience.

Sometimes, the poor villagers face their economic crisis in different situations. Therefore, they find out the way to solve their problems like–

- Poor villagers lease their land to neighbours in different situation like at the time of marriage of their daughter, treatment of their diseases, study of their children, etc. They get Rs. 6,000/-to Rs. 8,000/-per bigha for one year on the basis of fertility condition of their land and demand of both sides. The money lenders cultivate this land once in a year as an interest of total given amount. If the land owner does not to return back the money in due time then money lender will again cultivate the land.
- Those who are very poor in economy and have no cattle, poultry birds but have some capacity to rare, they take cow, she goat, ewe, hen, etc. from their neighbours and keep for a period on contract. Generally after the first issue of the said cattle/poultry bird, the rearer get the younglings. After second issue he/she returns the said cattle/poultry birds with younglings of second issue. This system is locally called as *Poshani*.

However, tourists are coming at Pakhirala village and go to the Sajnekhali Forest office about 1km. by boat and taking permit for visiting tourist spot of Sundarban. In 1984, West Bengal Tourism Development corporation set-up a tourist lodge at Sajnekhali and organized conducted tour to Sundarban. After that, private hotels/lodges were set-up at Pakhirala village and some private sectors also organize conducted tour from Pakhirala village to visit the Sundarban.

Due to tourism, local people are more or less effected. Therefore it is essential to know about the impact of tourism on local people. We have collected opinions about tourism of 300 individuals on both sexes of different communities. Out of which 71 per cent people favour tourism and 29 per cent people against tourism. The reasons for favour on tourism are as follows-

- During tourist season (October-February) hotel owners, boatman, tourist guide, petty shop keepers, van rickshaw puller are benefited as they earn more money. Beside this those who sell their vegetables, fruits, honey, fishes/crabs, etc., in the local market they are also economically benefited.
- Road, phone and transport facilities are developed.
- They are introduced to outsiders.

On the other hand some villagers put their opinion against tourism due to following reasons-

- As hotels/lodges are situated within the village the culture of the villagers are effected by the tourists. It is observed and reported that in the evening when some cultural programmes are arranged in hotels/lodges few tourists are engaged in drinking liquor and encourage dancing and singing which make a great noise and effect the culture of local people. Dress pattern, food habits of local people are also affected due to tourism.
- The cost of food items are increased in tourist season.
- The cost of land near tourist spot is also increased due to tourism.
- Tourist motor boat, vocal and instrumental music causes sound pollution.
- Outflow of diesels of motor boat causes pollution in river.
- Few tourists do latrine and urinate here and this causes pollution of local environment.

Now, it is essential to know about the Biosphere Reserve and its impact on the local people. Biosphere Reserve is the protected area of particular environment for conservation of biological diversity through sustainable livelihood which maintains the ecological balance. In 1989, Sundarban has been declared as Biosphere Reserve which is divided into following zones-core, buffer, and sanctuary and transition zone.

- *Core zone* is protected for conservation of bio-diversity. The people can not perform any kind of economic activities in this zone.
- *Buffer zone* is surrounded by core and sanctuary area and situated near Bangladesh border. This zone is utilized for fishing, collection of honey, wax, etc., with permission from the Forest department. Tourists are also allowed in this zone.
- *Sanctuary* is adjoining the buffer zone. It can not be utilized for fishing and honey collection. But tourists are allowed in some spots of this zone.
- *Transition zone* is outer most zone of the Biosphere Reserve. This zone is mainly used for settlements, agricultural activities and other economic activities.

Due to these rules and regulations, fishing/catching of prawn-seed/crab, honey collection are prohibited in the Sanctuary and core zone. Regarding fishing, if one person has Boat License Certificate then three to four members can go with him as a team for fishing. If the fishermen enter into the core area they are apprehended by the staff of forest office and they have to pay fine. If this incident is repeated then they have to pay more amounts to the forest official as a fine. The pass for fishing is not issued during the month of Chaitra (March-April) to Baisakh (April-May) because fishes produce spawn at that time. The pass for fishing is issued from the month of Jaistha (May-June). They use different types of fishing net/trap/hook for fishing. Both male and female use thick nylon net for catching prawn–seed in large number which damage the seedlings of prawn and also destroy other spawns which may damage the ecosystem. Catching of prawn-seed/crab in the river side damage the base of river embankment.

In case of honey collection the pass is issued by the Forest office once in a year *i.e.,* during middle of the month of Chaitra (March—April) to middle of the month of Baisakh (April—May). After honey collection, they separate the honey and wax and sell in the forest office at Government rates.

Due to the rules and regulations in Biosphere Reserve, economically villagers are not fully dependant on forest and river based resources. Collection of fuel wood is another problem for them. Previously they collected fuel wood from the forest. But at present, branches and leaves of different trees are collected from their surrounding of houses and used as fuel. Floating leaves of mangrove plants are also collected from river and used as fuel. Sometimes, they also purchase fuel wood from market.

However, mangrove plantations are done along the bank of rivers by the Social Forest Office and Eco–Development Committee. The Non Governmental Organizations are also working here to develop the area.

Regarding Biosphere Reserve, we have collected opinions of 300 individuals of both sexes of all communities in the study village, out of which 69.66 per cent people favour Biosphere Reserve and 30.33 per cent are against it. The reasons in favour of biosphere reserve are as follows–

- The forests protect them from natural calamity as well as preserve surrounding nature for their better life.
- Rainfall also depends on forests.
- They know about forest ecology and protection of flora and fauna in natural ways.
- Forest will be protected from poachers and wood cutters.

On the other hand, the reasons against Biosphere Reserve are as follows–

- Fishing and honey collection are allowed only in the buffer zone which is far away from their settlement. They lose their time in long journey to reach the permitted area *i.e.,* buffer zone.

- Due to long journey to reach the buffer area they get minimum time to catch fish/prawn-seed/crab which is not sufficient for their livelihood.
- Small boat is not sufficient to cross the long distance as well as in stormy weather. During natural calamity if they enter the sanctuary and core area for their safety and at that time if they are apprehended by the forest officials they will pay fine.
- The buffer region is situated near border of Bangladesh. Sometimes dacoits/antisocial persons attack them and seize their fishes, fishing nets and also demand cash money. These are actual problems faced by them in their economic livelihood.
- The people of Bagdi community generally depend on fishing activities. Some people of other communities are also depending on fishing. Due to these restrictions in Biosphere Reserve they are facing many problems in fishing activities in day to day life.

However, rules and regulations of Biosphere Reserve give rise to problems in the forest and river based economy, like-fishing/ catching of prawn seeds/crabs and honey collection. Therefore, rules and regulations of Biosphere Reserve are a barrier to their economic life and people are living in hand–to–mouth condition.

- Moreover, Govt. implemented many developmental and income generating schemes through self help groups, like-poultry, duckery, piggery, goat rearing, and manufacturing of fishing traps, etc. But these developmental schemes in most of the cases are not successful due to lack of proper training in regards to their culture and infrastructural facilities.

Discussion

It may be stated that earlier people of this village mainly depended on agriculture and local resources *i.e.*, fishing, collecting prawn seed and crab, honey collection, wood collection, etc. But their present generation has taken up different economic activities along with traditional occupation. Some of them have changed their traditional occupation to other economic activities due to impact of Biosphere Reserve, tourism and natural calamity.

Tourism creates some new type of economic activities for the people in various sectors, such as hotel business, tourist guide, boat-men and simultaneously increase the income of petty businessmen/shopkeeper, van-rickshaw puller, etc. Some individuals are also engaged in batik work and *dhadra* work (embroidery work) which may be properly marketed to attract tourist. Tourism also increases the income from pass (permit) to enter into tourist spot in forest area.

After declaration of the Sundarban as Biosphere Reserve some ban and restrictions have been imposed on forest resources which compelled the people to depend on alternate livelihood. Simultaneously due to excess fishing/catching of prawn seed, the availability of fishes/prawn seeds is less in number.

Agricultural production has decreased due to natural calamity, salinity, improper irrigation and destructive insects.

There is no such industry to provide jobs for the people. Therefore, seasonally 3.72 per cent people of total population have to go to Kolkata and adjacent districts like—Medinipur, South and North 24 parganas, Burdwan, Hoogly, etc. of West Bengal and worked as agricultural labour, daily labour, private job, etc. 0.78 per cent people of the total population go to Delhi, Bangalore, Mumbai, Tripura, Andaman etc. and are engaged in private job, daily labour, mason work etc.

Here, some points for benefit of the local people and tourists are mentioned below–

- Permission may be given for short cut route to go to the buffer area for fishing and honey collection in short time.
- Fishermen may use wider gap of fishing net instead of thick nylon net to protect spawns of fishes for ecological balance in river.
- Mobile phone, rubber tube and life jacket may be provided in the tourist boat for safety of tourist during dangerous situation.

- Electricity may be provided.
- Ferry ghat may be improved.
- Moreover, implement awareness programme about their economy, health, education, family welfare and environment through mass campaigning and audio-visual programme.

However, the people of all communities have to adjust in new situation for their survival. The people should not get restricted to their traditional occupation and males and females should engage in different types of economic activities for their better life.

References

De Barun (ed.)., 1994, *West Bengal District Gazetteers,* 24 Parganas. Govt. of West Bengal, Calcutta.

Hunter, W.W., 1998, *A Statistical Account of Bengal : Sundarban,* Vol. I, Part II.

West Bengal District Gazetteers (Reprint), Govt of West Bengal, Calcutta.

7

Human Element in Conservation Process: A Rethinking in the Context of Biosphere Reserve

Ratna Dhar

Introducing the Subject

Two key phrases often emerge in addressing the conservation issues. One the conservation process itself and second the role of human beings in the conservation process. The conservation initiatives around the world are separated into these two distinct stages of development. The early conservation literatures, more aptly known as Exclusionary approach regarded the local people as threats to conservation practices and the main purpose of establishing the protected areas were to protect the endangered species. After the 1980's there was certain shift towards the perception of the conservation among the policy framers when man was included in the conservation framework. Though it began with the formulation of Man and Biosphere Programme in 1972, the shift really took off with the coming of different conventions namely the World Conservation Strategy 1980, the Bruntland Report 1987, Biodiversity Convention 1992 and the Seville Strategy1996 (Bridgewater, not dated) as a concept of Peoples Approach (Ursula Grant, 1998). The main purpose of the shift in the policy was to give a large share of control over the conservation process to the local people.

The share of control brings out another dimension of integrating the conservation decision and the control over the development decision. This brings two sets of arguments between two polar groups. Using the coinage of Raymond Dasmann, the argument is between the 'Biosphere People' and the 'Ecosystem people'. The main argument of the Biosphere people who speak on behalf of the biodiversity is that the population living in or beside a reserve area and their economic activities are a threat to the 'carrying capacity' of an ecosystem and hold the idea that the local people are the main cause of degradation. Therefore the legal acts and policy formulations should be on the side of protection managers. On the other hand, there is another view that the human beings living in the center of the biodiversity cannot be ignored. For any kind of decision-making process in an ecosystem, the biological information and socio-political information are important which is held by local Ecosystem People.

At the policy level the goal of integrated conservation and development projects is to "enhance biodiversity conservation through approaches that attempt to address the needs, constraints and opportunities of local people" (Wells and Brandon, 1993) by involving local people as active partners. But it is found that between the two groups, the main bone of contention remains at the decision level where always the ecosystem person remains at the receiving end.

In 1988, with the growing priority on the conservation, the National Forest Policy also mooted the idea that the local communities are to be involved in natural resources conservation along with an emphasis on meeting the subsistence requirements of the people. Subsequently, the Indian Ministry of Environment and Forests issued a circular in 1990 for the joint forest management (JFM) and resource sharing. The user groups receive usufruct rights only and the land is not allocated or leased. To the protection managers or forest officials the JFM is primarily a means to ensure the rehabilitation of degraded forests but to the villagers it was a solution to the growing shortage of biomass. The protected areas in India in most of the cases are not situated in isolation due to human dominated rural landscapes in and around them who derive the resources of the protected area for their subsistence. The India Ecodevelopment Project was conceived as a pilot project in 1994 as a participatory management strategy of protected areas on the basis of an Indicative Plan prepared by the Indian Institute of Public Administration on behalf of the Govt. of India after the study of eight sites selected by the Ministry of Environment and Forests.

The eco-development aims at conserving the biodiversity by addressing both the impact of local people on the protected areas and the impact of the protected areas on local people. The scheme lays stress on implementation of the measures that are most suitable from conserving the biodiversity. These measures are divided in two parts. Firstly, those works which improve the natural environments in the protected area and generate employment for the local people, and secondly those that generate alternative sources of income for the local people and change their dependence on the protected area. As a strategy for implementation of the ecodevelopment scheme, the buffer zone are managed mainly with the objectives to prevent the wildlife living in the core zone to interact with the human population and provide them a natural surrounding and to meet the minimum needs of the local population for forest produce and for the cattle-grazing, collecting dead woods etc. In most of the areas, the following activities and works are undertaken under the ecodevelopment scheme to meet the requirements of the area as:

(a) To make available green fodder (b) Development of pastures (c) To make available of fuel wood (d) Soil-conservation work (e) Water conservation work (f) Fruit bearing plantation (g) Pisciculture (h) Apiculture (i) Development of cottage industries (j) Development of eco-tourism (k) Promotion of sources of alternative energy (l) Animal husbandry. It is found that in many cases these activities overlap the multiple zones of the buffer and also to the transition areas on the community and private lands.

The present article examines the populist approaches of biodiversity conservation and ecodevelopment programmes implemented in the fringe villages of the Sundarban Tiger Reserve.

The focus is in one village and the livelihood strategies employed by the resident villagers in this deltaic mangrove zone.

Sundarban Tiger Reserve

The Sundarban Tiger Reserve was constituted in 1973 and comprises 2585 sq.kms and forms a large part of Sundarban Biosphere Reserve, which is about 9630 sq.kms in the Indian part. The Sundarban Tiger Reserve (STR) is bounded in the east by the international boundary with Bangladesh formed by the rivers Harinbhanga, Raimangal, and Kalindi. The Bay of Bengal is situated on the south. The western border is along the Matla rivers thereby form the boundary with territorial Forest Division of 24 Parganas (South) and Northwest is bounded by river Bidya, Gomdi, Kapura, Korankhali and the Kalindi (Annual Report 2003-04). These rivers form the boundary with the revenue villages bordering the Tiger Project. Being a larger part of Sundarban Biosphere Reserve, the Sundarban Tiger Reserve includes the Core zone of 1330 sq km and the Buffer zone of forest cover of 1255 sq km of mangrove forest. The core area includes the Sundarban National Park (1330 sq.km.) and is fully protected by the Wildlife Protection Act, 1972 and Forest Conservation Act, 1980 and strictly protected from all development activities. The Forest Conservation Act, 1980, gives the legal protection to the Buffer zone, which has three sanctuaries namely, Sajnekhali (362 sq.km), Lothian Island (38 sq. km.) and Haliday Island (6 sq. km.). The tourism activity in this part of the biosphere is mainly controlled by the Department of Forest with the spots located at Sajnekhali, Sudhanyakhali, Netidhopani, Dobanki, Burirdabri, Bonie Camp, Bhagabatpur Crocodile Project, Lothian Island Sanctuary and Kalash Beach.

The revenue villages situated on the fringes of the tiger project are the last human settlement of the area and forms the part of transition zone of Sundarban Biosphere Reserve and stretches to 5367 sq. km area which includes13 blocks of South 24 Parganas district (Sagar, Namkhana, Kakdwip, Patharpratima, Kultali, Mathurapur I and II Jaynagar I and II Canning I and II, Basanti and Gosaba) and 6 blocks of North 24 Parganas district (Hingalganj, Hasnabad, Haroa, Sandeshkhali I and II, and Minakhan.

Sundarban Biosphere Map

The following Box I explains the situation:

Box I Biosphere Zonation and Protection

Zone	*Area*	*Forest Dept Activity Zone*	*Tourism Zone*	*Legal Protection*	*Traditional Resource Zone*
Core	1330 Sq kms	Part of Sundarban Tiger Reserve. Sundarban National Park	— 1972,	Wildlife Protection Act, cutting Forest Conservation Act, 1980.	Fishing Wood
Buffer	1255 Sq kms	Part of Sundarban Tiger Reserve. Restoration of mangrove vegetation 3 sanctuaries Sajnekhali Lothian Island Haliday Island	Sajnekhali Burir dabri Dobanki Netidhopani Sajnekhali Tourist Lodge	Forest Conservation Act, 1980	Fishing. Wood Cutting. Honey Collection
Transition	5367 Sq kms	Eco Development Programmes Zone of Cooperation	Government and Private Tourist Lodges		—Villages, Human Settlement. Used for agriculture, fishing, Use of local ecosystem resources.

The ecodevelopment activities were introduced in 25 fringe villages (22 villages in the Gosaba block and 3 villages in Hingalgunj block) of the Sundarban Tiger Reserve (STR) in 1998. This was done to reduce the peoples dependence on natural resources with the adoption of the ecodevelopment programme formulated through micro planning at the grass root level on the basis of need based assessment adopting a method of Participatory Rural Appraisal (PRA) involving the people and taking the available resources into consideration. In the STR zones this programme was thought to be a solution through which the forest would be protected, people's dependency on the forest would be lessened and the long term sustainable development would be done. The approach was to come to people and give some work to them. Cooperation between the two would bring a substantial balance in the total environment.

Map of Fringe Village

Box 2 Eco Development Programmes in STR

In STR zone, the real problem with the forest department and the government agencies is the protection of the whole area as the core area is threatened by the population pressures in transition zones which results in the collection of timber, fishing, & other forest conditions. In the buffer zone illegal felling, woodcutting, poaching of firewood, and tiger prawn seed collection are the resultant effects of population explosion and the encroachment. Illegal felling of trees, monoculture technique of prawn culture and conversion of agricultural land to prawn culture are the adverse activities on the transition area. On the other hand, the villagers living in this ecozone face the straying of tiger in the habitation, river erosion, and storms and cyclones, large-scale expansion of aquaculture and prawn farms and loss of productivity of the cultivable land due to salinity for years together.

Study Village

The Dayapur village in the fringes of tiger project reserve was selected for microlevel coverage. It is one of those 25 villages where ecodevelopment work is promoted. The village is situated in the Satjelia Island under Satjelia Gram Panchayat in the Gosaba block. The Gomor river bounds the village on the north. On the north of Gomor river is the Gosaba island where the block headquarter is situated. The nearest village in the Gosaba island is Pakhirala where the Sajnekhali Forest Range office is situated. To the south is the river Pichkhali. On the southern bank of the Pichkhali river, is situated the Sajnekhali Island where the Sajnekhali Wildlife Sanctuary is located along with the Mangrove Interpretation Centre and Sajnekhali Beat Office. To the east is situated the Sukumari village and the western side forms the conjunction of the two rivers. The village is approachable from Gosaba, which is also the nearest market and can be reached by travelling in the cycle van for about 10 km and then by *bhot-bhoti* in river Gomor for about 10 minutes.

The origin and growth of the village is a part of reclamation work in the jungles on the northern and on the southwestern part of Sundarban taken up by the British administration in different phases in between 1873-1939. In 1903 the Zamindari area of 9000 acres of Gosaba Island belonging to 142 and149 no Lat was taken by Sir Daniel Hamilton. Later as a part of extention of zamindari in 1920 Sir Daniel Hamilton took the zamindari of Satjelia island about 13000 acres and reclaimed it for human habitation.

Population and Growth

The first entry of the people began in this part of Sundarban from 1903 to 1909-1910 in the midst of reclamation. In 1912-1913, 300 tribals belonging to Santal, Munda, Oraon and Bhumij groups came from Purulia. They first stayed at Sandeshkhali and then went on to Dayapur, and Satjelia (Jatirampur Ghat). The tribals as cheap labour (indenture labour) were possible due to the middleman (*arkathia*) and *girmite* (a localized form of agreement in Chotanagpur). As a part of the contract, the tribals got the land for their subsistence, but most of them could not keep the control. These middlemen in due course grabbed the lands of the tribals. At present, the tribals are mainly concentrated nearer to the Sundarban Reserve forest area. The largest concentration is found in Gosaba PS where the population strength is around 20,000. Around 1930-1935, about 1000 peopl from Midnapur and 24 Parganas came. They settled in Baly II and Palpur, Bipradaspur, Gosaba, I. Rangabelia, Amtali, Sandeshkhali, Joygopalpur, Korabati, and Sukhdoania and acquired lands here. The Adibashi population came here around 1917. In 1948-49 around 50-60 families came as refugees to Dayapur. About that time 20 Muslim families were residing in the village on the Khas land. They left the place by exchanging the land. Some families started living near the riverside nearer to Binapani F.P. School. About 500 hundred acres of land originally belonged to the tribals. After the dissolution of Zamindari Act, the refugees from East Pakistan came to Dayapur in 1958. 350 families came and were given 5 Bigha of agricultural land. In 1962-63, 300 families came to Dayapur and settled here. In the midst of riots in 1964-66, 4-5 families came and settled to the south of Binapani. Within four years 400 families settled here. After that every year about 10 families are coming to Dayapur barring the last 3-4 years as restrictions have been imposed and strict vigile is kept on the borders. The people are mostly from Khulna district and few have come from Faridpur and Bagherhat. In the last few years, the population has increased in Lahiripur, Amtali, Satjelia, Kachukhali, Mollakhali, Dayapur, Kumirmari and Radhanagar. Muslim population has increased in Basanti, Sonakhali and in the direction of Canning. There was always the population influx from Bangladesh but in the recent years the people coming from Bangladesh are not settling at the islands. [1]

Presently Dayapur village has heterogeneous populations. It has as many as eight Hindu castes and two tribal groups. The two tribal groups Bhumij and Munda have lost their tribal identity almost completely and have become part of the local peasantry. From the Table I, it appears that 914 families are residing in the village. Pondro Khatriya is the single largest group of the village constituting 755 families followed by the Bhumij who comprises of 130 families. All other inhabiting groups of Dayapur have relatively very small populations.

Table 1 : Family wise Distribution of Ethnic groups

Total	*Caste Groups*	*Tribal Groups*
914	781	133
100	85.45%	14.55%

Table 1.1 : Caste Groups (Family wise)

Total	*Brahman*	*Kayastha*	*Mahishya*	*Jugi*	*Kapali*	*Napit*	*Pon Khat*	*Muchi*
781	10	2	6	1	1	4	755	2
100	1.28%	0.26%	0.77%	0.13%	0.13%	0.51%	96.67%	0.26%

Table 1.2 : Tribal Groups (Family wise)

Total	*Bhumij*	*Munda*
133	130	3
100	97.74%	2.26%

Table 2 : Age Group wise Population Distribution

Age Groups in years.	*Male*	*Female*
0—14	610	626
15—59	1431	1312
70—+	62	34
Total	2211	2063
%	51.73	48.26

A distance of one km beginning from the forest on the side of Pichkhali river and Gomor of the village around the riverside is demarcated under EDC which includes the whole of Renuka Nagar Colony, Baghedhara, the Topor para hamlet on both sides of Topor Khal and part of the Mridhapara-Mazherpara hamlet on both sides of Malshabhanga Khal. The following Table shows the land use of the existing village data.

Table 3 : Land Use Dayapur 2006

Sl. No.	*Land Characteristics*	*Area in Cents (Percentage)*
1.	Arable Land	65.1
2.	Residential Area	14.2
3.	Water Bodies	7.9
4.	Eroded Land	5.3
5.	Tourist Lodge	0.05
6.	Others	7. 45
		100

The village data shows that the arable land is 65.1 percent of the total area and the residential area is 14.2 percent. This is due to the population influx in different periods in this village. The residential area has spread to many parts of the agricultural land and has encroached the agricultural area. The residential area has also spread to both the sides of the four canals of the village, *i.e.*, Topor Khal, Malshabhanga Khal, Abirkhali Khal and Kumirdoba Khal.

The present area covered by the water bodies is 7.9 per cent and it is found that many of the intricate water canals which earlier existed are now nonexistent due to silt formation and included in the agricultural land.

The villagers regularly face the embankment erosion, which increases more during the rainy season. At present the eroded land has covered 5.3 per cent of the village land area. On 10.6.1999 (26 Jaistha, 1406 Bangada) in the Baghedhara and in part of Renuka Nagar area on the southeastern side of the village a large portion of the riverbank of Pichkhali river suddenly eroded and a large flow of water about 4000ft entered the area. A large area of agriculture and the residential land belonging to inhabitants of Baghedhara, Renuka Nagar and Mazherpara was lost in it. The Major part of the village area was flooded with water. For 15 days the ebb and high tide flew in the landed area of the village. It took 45 days to repair the ring embankments. The original land where the Sundarban Tiger camp resort was to be established was denuded by the river Pichkhali and shifted to another nearby land after an embankment was made. As a result a part of the residential and agricultural land is now turned to *Veri* (fishery farm), which is the largest in the village. Many of the residents of Baghedhara para have shifted to Banikhali. The water bodies and broken remnants still show the evidence. The salinity increases in the lands with the silt coming with the high tide and it takes 1 year to remove the salinity. The second veri is situated to the west of the Sundarban Green House. This is also the result of the erosion in which the agricultural lands of the residents of Renuka Nagar Colony were lost.

In September 2005, the high tide and a large area of the agricultural land denuded the western part of the village and one fishery was completely under saline water. The saline water also entered in the Malshabhanga canal. The village has taken up the social forestry programmes by the EDC and mangrove (Baen) saplings were planted on the riverbanks to check the river erosion. The river erosion has added many problems to the villagers. They are loosing the agricultural land and the investment made for the agriculture is lost. The villagers often complain that they are unable to utilize the land that is lost in erosion but are paying tax for it.

The village has three privately owned hotels, which engage in tourism activity and covers 0.05 per cent of the land area. The only NGO of the village Sundarban Institute of Alternate Income Generation is situated nearby the Sundarban Green House. This NGO mainly works in nearby villages in the Self Help Group of the eco development programme. In recent years it has also started accommodating the tourists in the institute.

Traditional Economy

The people who came to this ecozone were dependent on the resources available in the land, forest and in river or sea for their survival. Though they were traditionally dependent on the agriculture, the area also offered resources in the dense forest and in the high tidal rivers to eke out a living. The following chart gives a brief account of the category of resources traditionally available to them.

Traditional Resource Base

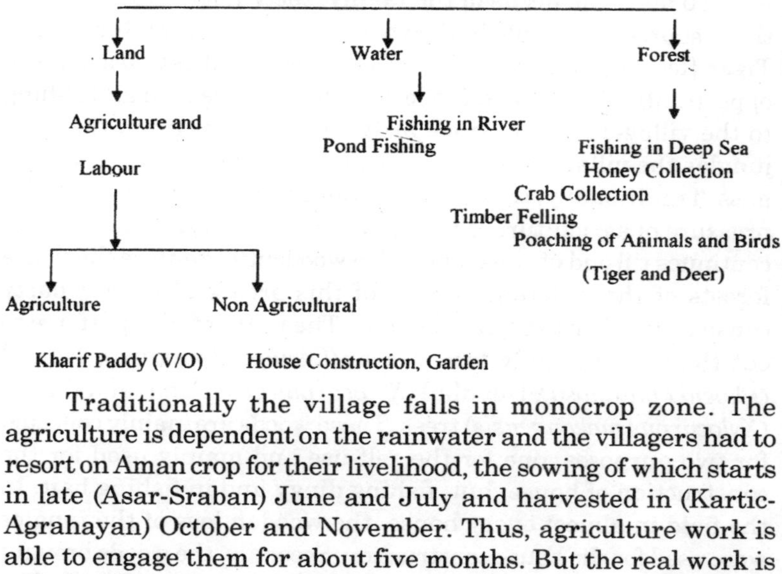

Traditionally the village falls in monocrop zone. The agriculture is dependent on the rainwater and the villagers had to resort on Aman crop for their livelihood, the sowing of which starts in late (Asar-Sraban) June and July and harvested in (Kartic-Agrahayan) October and November. Thus, agriculture work is able to engage them for about five months. But the real work is only for tilling the land (fifteen days), sowing (fifteen days to one month), weeding (fifteen days) and harvesting (one month). In the Kharif season, labour is needed for sowing of saplings in Ashar and Sraban. Women are mostly engaged in sowing. The weeding season again brings some work for about 15 days though the weeding (*khoyali*) is done from time to time. The women and elderly males mostly engage in the weeding process in the village. Thus, in real sense the agricultural season could provide work to a villager for about two and half months regularly. So to meet the yearlong demands of the family and living, a certain percentage of the males of the village use to go to work in Hoogly, Barddhaman for sowing and transplantation in Ashar and Sraban after completing the transplantation at home. The males remain outside for work from Sraban to Kartic and also in Agrahayan till the harvesting (*dhan kata, dhan tola*) is over. They usually returne home during Durga puja and return to the work place after Lakshmi puja. Traditionally the non-agricultural works available in the village are the construction of thatched house, gardening or digging of ponds. But the monocrop agriculture and the seasonal agricultural labour are not sufficient for the village, which faces the population influx day by day.

To meet the needs of the family, the villagers also opte for the resources available in the vast reserve forests of Sundarban Tiger Reserve. The reserve forests, rivers and estuaries offere opportunities for the wood collection, honey collection and fishing to the villagers of the fringe areas. But to venture in the deep jungles the villagers has to seek a token permission and a BLC pass. Traditionally, the season of timber felling begins when the pressure of agriculture is negligible from the mid of August and continues till end of November. The woodcutters entere the dense forests of the different islands of this area and form a party consisting of five to fifteen persons. The timber feller preferre to cut the woods mainly from Garan*(Ceriops decandra)*, Hental *(Phoenix paludosa)*, Dhudhul (*Xylocarpus granatum*) and Passur *(Xylocarpus mekongensis)* trees. These woods are mainly collected for two purposes, one for the self-use and mainly used for the construction of house, hut, fishing dingy and in fishing bait. In the field in almost every house, the wooden logs of the jungles were used for the house construction. Secondly, the woods brought from the jungle were sold in open market in the village. The villagers even used the woods for the fuel purpose. The demand increases in Poush and Falgun as the season is dry and the villager is able to use his man-days in the construction work.

Honey is available in the forest from the late Chaitra (April-May) and continues till Jaistha (May-June). The villagers take this opportunity and form a group and the collection of honey is done in a very disciplined way. A group of 7-8 people make a line and cover an area of 5-6 *katha* for collection. A fire is made so that the honeybees may fly away. While cutting the comb one person takes *ari (dhama)* underneath the comb so that the honey drops in it. The *dhama* is then taken to the dinghy where the honey and wax are separated. After the return to the village the honey and the wax is sold. The forest department has restricted the collection of the honey.

The hunt of tiger was also associated with the indigenous knowledge of the people. It was believed that a small pinch of the tongue of tiger causes infertility. The moustache is useful to cure ear pain. Lips are pasted to powder and when mixed with honey and smeared on the neck and throat, it cures the fracture or any injury of throat. The flesh when boiled in oil and smeared on body heals up any kind of body pain.

The cocks are brought from the jungle which are left in the jungle for some vow and brought from jungle mainly for cockfighting (*morogkhela*) which has become a pastime for villagers of the area.

The villagers mainly go to the estuarine area for two reasons, one for fishing and two for catching the crabs. Traditionally fishing is done by two methods; the first method is by fixing nets (special kinds of nets are used for particular type of catch) in strategic areas for catching the large fishes. The second method is the deep-sea fishing.

There are again two ways by which a fishing team enters the river and estuarine area. The first one is the longer one and a team takes 6 days to make a full circle of the riverine area. The team members may be 5 or 6 persons and take the following items for going to the forest—dinghy 2 or 3 in nos, dar, boitha, cooking arrangements, ration for 8 days (water, rice, vegetables, masala, oil, salt.), don, borshi, dry fishing bait (sukno chara), crab, hungar macher chara, live fish (*chingri, parshe*), BLC pass, (Mach sukno kare *chara of erals* banano hoy.), *jal* (net). The journey usually begin during *bhata* time on *dashami* or *ekadshi* and return in *jowar* by Sashthi or Saptami.

The second one is comparatively of short duration and in every visit to jungle they leave the home early morning in a group of 3 to 4 persons in a dinghy. The small fishing boat (*dinghy*) is useful while going to collect fish, crab, and prawn because these can be taken to the creeks where the catches are found in plenty.

Though the crabs are available throughout the year but in the seaside area the crabs are found more during the summer. The number of crab increases in Aswin and Kartic and by the end of Falgun the number decreases. During the months of availability, the number increases during the *gon* days mostly. The increase in the number is found for continuously three days before and after Amabashya and Purnimas. The crab collectors form a group of 2-3 persons and go together in the small rivers and creeks dinghy before sunrise when the jowar starts, catch the crabs and return by afternoon in next jowar. They take the *jalti* (net used to catch crabs), bait, *haari* (pot) for keeping the crabs, etc.

Thus, traditionally calendar wise the following resources are available to the villagers for their survival. The search for the honey or logging of woods or fishing brings them to the real crisis of the life of directly facing the death and different hazards of the deep forest that includ attacks from the royal Bengal tiger, crocodiles and other wild animals and snakes.

Box 3 : Calendar-wise Resource Availability

Bengali Month	*English Month*	*Resource Availability*
Baisakh	Apri-May	Honey
Jaishtha	May-June	Honey
Ashar	June-July	Sowing of Aman crop
Shraban	July-August	Sowing of Aman crop
Bhadra	August-September	Weeding, Wood felling
Ashwin	September-October	Wood felling, Crab collection
Kartic	October-November	Harvesting of Aman crop, Wood felling, Crab collection
Agrahayan	November-December	Harvesting of Aman crop
Poush	December-January	Hut making, Thatching, Tank digging
Magh	January-February	Hut making, Thatching, Tank digging
Falgun	February-March	Hut making, Thatching, Tank digging
Chaitra	March-April	Honey

*Fishing throughout the year

The occupational hazards are always accompanied by uncertainty of life and inevitable risk, hence the poor villagers have developed a symbolic channel of communication with supernatural powers and the belief systems connected with the propitiation of the tiger associated deities (Niyogi, 1996). The fear of tiger is not only in the jungles but many a time tigers stray into the villages in search of prey that destroy not only the human lives but also the crops in the agricultural lands.

Formation of Sundarban Tiger Project and Economic Impact

The villagers are able to meet their two ends meal and occupation in the same way till 1973 when 2585 sq.km forest area of Sundarban came under the Tiger Project on a 'core-buffer strategy'. Strategically the core area is kept free from the biotic disturbances and forestry operations, where collection of minor forest produce, grazing, human disturbances are not allowed. But the buffer zone is managed as a 'multiple use area' with twin objectives of providing habitat supplement to the spill over population of wild animals from the core conservation unit, and to provide site-specific ecodevelopmental inputs to surrounding villages for relieving the impact on the core area. Non-Timber Forest Produce (NTFP) collection and other rights and concessions to the indigenous communities are permitted in a regulated manner to complement the initiatives in the core unit. (www.envfor.nic.in)

The immediate fall out of the programme is that the whole forest area of 2585 sq kms is now squeezed. The villagers of the fringe areas are restricted from entering into the core area of 1330 sq kms, though they could use certain areas of buffer zone. Thus, the vast spread of area that was earlier available to the people for their economic pursuit is now restricted. This means more human impact on a less area. The restrictions also create legal problems for them.

The restrictions imposed by the formation of Project Tiger forced a good number of populations to switch over to a new kind of occupation in the area in the form of collection of tiger prawn seed in early 80's. The Sundarban region has naturally available seeds of Tiger Prawn (*Penoeus monodon*) throughout the year in varying quantities. (Bhowmic 1992: 38-41; Chakroborty et.al 1977:181-183 c.f. Bhattacharya and Ray). In the month of March-

April when the wind blows from south, thousands of prawns enter into the different rivers and lay thousands of eggs.

The tiger prawns are collected by two methods, by using net with floats and by drag net. The dragnets are used at the beginning of the *bhata*. The catching of *bagda* begins during the onset of the low tide. Summer season is considered as the best time for collection of good amount of prawn seeds. The winter months are the poorest season for the collection of prawn seeds, but the price of *bagda* remains highest during winter. This may be because of high demand and low supply. During winter season the breeders need the seeds for aquaculture/fisheries.

People from other occupational groups such as cultivators, day labourers etc, temporarily engage themselves in prawn seed collection to get some ready cash. Next to forest this is the area where easy money is readily available and women and children could supplement the earnings. The unavailability of the alternate source of income in the village forces the *meen* (tiger prawn seeds) collectors to collect the seeds. But this has caused other problems. One, the area being in high active delta zone is prone to riverbank and embankment erosion due to the turbulent river. This leads to the entry of the saline water in the settlement area and in agricultural area. The process is again aggravated due to the continuous dragging of the nets along the riverbank. Secondly the fishes of the sea enter the riverine area and estuary for fertilization and produce the eggs. After the fertilization, a large part returns back to the sea but a portion moves towards the rivers, creeks, and rivulets. But due to fine netting these are destroyed and the fisherman who is not getting enough fish feels the ultimate impact.

Declaration of Biosphere and Impact in the Traditional Economy

The attention of the world community increased more due to the declaration of the area as global biosphere in 2001. This has increased more insecurity among the villagers. A common villager is ignorant of the demarcation of the territories of the biosphere and forest area. He knows from where he can procure the items of his need. In fact in the villages being situated on the periphery of the transition zone, there is every chance of crossing the territories of the buffer area.

Earlier timber license for the timber felling was given to the villagers, which has now been legally stopped for since 1998 by the Forest department for which the collection of timber has become difficult. The honey collection is continued as Non Timber Forest Product and involvement of Forest Department has increased. The fishing in river is continued with the proper permission from the Forest Department. Fishing is allowed in certain areas of buffer zone *i.e.,* in Arbeshi/Arbesi, Jhilla, Harinbhanga, Khatuajhuri, Chandkhali and Dutter border and prohibited in Baghmara—Compartment I, East Range—Compartment I-VIII, Gona—Compartment I-III, Champta—All Compartments, Sanctuary area—Range, National Park West Range, Netidhopani—Compartment I, II, III, and Haldibari—All Compartments falling in parts of buffer area and core area.

The biosphere restrictions have forced them to change the traditional route where they could get more fish in fishing pockets thereby loosing the traditional base. To the understanding of the fishermen the area has come under the core and buffer zone and the fishermen are asked to fish in the eastern parts of Goral river which is towards the Bangladesh border. At present fishing is not allowed on West of Goral river and east of Bidya river upto Bay of Bengal being in core area but these form the pockets of fishes. Now fishing is allowed in the areas east of Goral river and west of Bidya river. But the area, east of Goral river is dangerous because of repeated disturbances and violence of the Bangladeshi dacoits.

As a result, the work force engaged as day labour in the village and outside either as agriculture and non-agriculture worker have outnumbered (274+12+207+89=582) the agriculturist (400+38=438). In recent times after completing the transplantation of paddy at home in Ashar and Sravan as per requirement, the males go outside to work in Hoogly, Barddhaman for sowing and transplantation. The weeding season again brings some work for about 15 days though the weeding is done from time to time. The women and elderly males remain at home and engage in the weeding process in the village. The crop usually becomes ready for harvest in November or December. The males remain outside for work from Sraban to Kartic and also in Agrahayan till the harvesting (*dhan kata, dhan tola*) is over. They usually return home during Durga puja and return to the work place after Lakshmi puja.

A section of male workers are engaged in non-agricultural works in the village or in the nearby villages. The labourers are engaged during Kartic and Agrahayan for preparatory works in vegetable gardening (*bagan bari*). The months of Poush to Falgun are used for repairing works in the house which include the thatching, fencing, mud wall construction and construction of the house (*gharami*) in the villages. During this season the harvest is already at home, the threshing is almost complete, hence the raw materials such as straw are readily available at home and the season is also comparatively dry. A portion of the work force also remains outside to work for non-agricultural works in this season. Even the villagers engage labourers for digging ponds or clearing the waste materials or felling soil in the road at their own personal cost. A section of villagers work outside in Tripura to work for fitting of pipelines and tubewels from August to October. Some also go to Andaman, Kerala and Tamil Nadu. For outside work in agriculture as well as for construction and fitting, the Kolkata based contractors engage them through the agents.

The villagers now use the dried twigs of the trees, bamboo roots, tree branches and leaves of the deep jungles when they flow much inward in jowar period for fuel. They collect the jungle articles during the jowar when the water movement is more in the islands.

The biosphere restrictions thus have on the one hand force the villagers to adopt new avenues of source of income, moreover, the people have also found out the loopholes and adopted new ways to manipulate the restriction terms.

Many a time fishermen enter the restricted zone without informing the officials. When they go without permission, it is called black and the persons are identified as Blackmen. There are also cases when though the fishing party obtains the legal permission from the forest department for fishing in the permitted areas but also collects some quantity of wood and honey which is passed to the village at night through a known dinghy, or the fixed dinghy, which is earlier, informed to collect the wood. The whole process is done secretly.

Many villagers of the surrounding area go for timber cutting. They do it in a secret way without anyone's knowledge. Four to seven persons make a group along with *Dinghy, Dao*, and *Kurul* and go in the early morning at the beginning of *jowar*. They take *panta* with them for the day and work in the forest for 12 hours and return again when the next jowar starts. This is mainly done to earn money but for the own consumption the villagers are now depending on the available trees in the local market or using the trees grown in the courtyard.

The importance of Satjelia Bazaar in Satjelia village has increased among the villagers of neighbourhood. It is the place where a villager can get woods of his choice from the jungle. There has been increase in the strictness so now the smuggled woods are sold in Satjelia 2 no bazaar secretly. It is a chain system where the woods of the forest are brought illegally and change several hands for required purpose.

The occupational data available from the village shows the above picture.

Eco-development Programmes

In 1998, 316 households, (from one household two members *i.e.,* husband and wife) became members of the ecodevelopment programme as a part of the larger ecodevelopment programme in fringe villages of Tiger Reserve. By 2006, the membership has increased to 544 households. As earlier said the eco-development is an attempt to reduce forest dependence and to compensate local communities-in cash and kind as well as through alternative off-farm income-generating opportunities-for the lost access to resources in protected areas. The approach acknowledges that people living near protected areas may have to bear enormous opportunity costs while deriving few tangible benefits from conservation (FAO Document). Though the official beginning was in 1998 but the work started from 1996-96 with the restoration of Topor Khal and Pond excavation of P.C. Sen High School. The following development works were carried out since 1996.

Table 4

Infrastructure Development in the Dayapur and Community Development Projects

1996-97	Restoration of Topor Khal	820 mtrs
1996-97	Pond excavation in PC Sen High School	30x50 cu mts
1998-99	Brickpath construction Solar Lights-1	100 mtrs
1999-2000	Brickpath construction	413 mtrs
	Pond excavation in PC Sen High School	
	Social Forestry in Riverbanks (embankments)	
	Social Forestry in P.C.Sen School	
	Social Forestry in Pond Ghat	
2000-01	Brickpath construction Plantation in Riverbanks (Nadir char)	800 mtrs
2001-02	Restoration of Malshabhanga Canal	575 mts
	Solar Lights-1 (P.C.Sen High School)	
2002-03	Solar Lights-1 (P.C. Sen High School)	
2003-04	Restoration of Malshabhanga Canal	1075 mts
	Brickpath (Dayapur Bazaar to P.C. Sen High School)	268m
	Solar Light in Dayapur Bazaar	
	Distribution of Smokeless Chullah	150 nos
2004-05	Restoration of Malshabhanga Canal	664 mts
	Brick path construction 3x350 m	1050 mts
	Distribution of smokeless Chullah	120 nos
	Formation of Self Help Group	14 nos
	Construction of Brick Jetty	1 no
	Restoration of Irrigation Canal	1 no
2005-06	Brickpath construction	180 ft.

Community Development Programme

The above table shows six types of community development works were carried out in the village.

(i) Restoration of irrigation canal

(ii) Pond excavation

(iii) Social Forestry/Plantation

(iv) Solar Lights in public places

(v) Construction of Brick path

(vi) Construction of Brick Jetty

The restoration of irrigation canals was an attempt to join together with the forest officials for the betterment of the village. An effort was taken to clean the Topor Khal (irrigation canal) in1996-97 for the benefit of the villagers. The canal was choked with the silt of the river and was not fit for use. For restoration of the canal, labourers were needed which was available in the village itself. The restoration of the irrigation canal help in irrigating the agricultural fields, which further help in increasing the crop production. Now people could think of putting more effort in agriculture. This had led to the villagers to believe that a joint effort could bring some development for the villagers and the forest personnel's can help in giving some work and idea for their betterment. During 1996-97, 820 mtrs of Topor khal was restored. In 2001 the restoration of another canal of the village, Malshabhanga started which continues till date in different phases.

Now though almost every farmer of the village is engaged in Aman cultivation, but now a large number of people living near by the canals or those who have agricultural lands nearer to canals start Boro cultivation in winter, which requires irrigation. The agricultural lands are getting water for irrigation. The farmers are benefited and more cultivation is done. So those farmers who reside on two sides of Topor *khal* and Malshabhanga *khal* are cultivating the Boro paddy. The people residing nearer to Abir *khal* and Kastura *khal* have also started cultivating the Boro paddy.

The renovation of the irrigation canals is a turning point in the life of many villagers of the area. The farmers are benefited and more cultivation is done. With the renovation of the irrigation canals, double cropping has started in a limited way since1980. But though the Boro crop brings high yield, it also requires high yield seeds, fertilizers and water. So it is not possible for all to go for double cropping.

The village data on agricultural landholding shows 562 families dependent on Kharif cultivation along with other cultivation. Now with the restoration of the irrigation canals and the impetus given by the ecodevelopment committee, 168 families have started cultivating the Boro crop. Now the preparations of the land commence in Agrahayan (November) and Poush (December) when seedbeds are raised. Normally seedlings are sown in the field from seedbeds after a month. The transplantation begins from Poush (December-January) to Falgun (February-March). The male members go outside for agricultural work in Poush (December-January) and return by the end of Magh and Falgun (February-March). The harvesting commences from late March (*chaitra*) and continues to April (*baisakh*). In boro paddy, 10 kg of seeds in 1 bigha will yield 18–20 *maunds* of Paddy. This cultivation is expensive but the yield is high.

The restoration of the irrigation canal has brought another change. It was found during the field survey that almost all the villagers living near the canals were using the water of the canals in irrigating the Boro crop. When the water was reduced, nets were used to catch the fishes, which were subsequently sold in the market. In most of the cases the small pond near the agricultural field was used for irrigation. Then the women, young boys and girls go inside the pond to collect *chuno* (small fish) for the consumption.

The houses on both sides of the khals have the fishing shares of their own and utilize the irrigation canal for fishing. For the last 8-10 years, the Panchayat is leasing out the canals to the people for fishing in a cooperative basis. All the important canals of the village *i.e.*, Malshabhanga, Topor, Abir, Kastura have been leased out to the villagers jointly by the gram panchayat. The houses on both sides of the khals have the fishing shares of their own and utilize the irrigation canals for fishing.

The pond of Dayapur P.C Sen High School was excavated in two phases, in 1996-97 and in1999-2000 and the villagers are able to use the water for drinking.

Social forestry was mainly done in areas, which was prone to erosion or where a great part of the residential and agricultural land area is washed away to check the erosion.

During the rainy season most part of the village path remains either water logged or is under water. It is very difficult for the villagers' to move in different parts due to the clayey soil. The importance of brickpath was felt for easy movement the construction of which began in 1998-99 by the ecodevelopment committee, though the Sundarban Development Board and the Panchayat has also contributed its role in brickpath construction in different parts of the village. By 2006, a large part of the village path was brick laid.

Individual Beneficiary Programme

The distribution of smokeless chullah and formation of self-help group was confined to individual family level.

The trainings were imparted under Self Help Group in hen rearing, duckery, goatery and piggery for this village, which were arranged by the Forest department. The Forest department took the help of NGO's in Training and in follow up programme. The thrust of Self Help Group in this village has been to the areas nearer to the river that are thought to be more dependent on the forest. In 2005, 14 teams were included in the Self Help Group. Five programmes were introduced. Piggery was one of them, which was vehemently opposed by the villagers, even by the tribal populations, as the enviroment would be dirty. So four programmes were taken. The beneficiary to whom piggery was allotted were later included in the goatery programme.

Box 4 : Self Help Programme in Dayapur

Self Help Programme	*Part of the village*	*Ethnic Group*	*Result*
Poultry	Sides of Malsabhanga Canal	Non Tribal and Tribal	Continuing
Duckery	River Side	Non Tribal	Failed
Piggery	Pockets of Tribal residence	Tribals	Did not begin
Goatery (4)	Topor Canal	Non Tribal	Good Result
Paddy Husking	Malsha Bhanga and Renuka Nagar	Non Tribal	Poor

Impact

Poultry—To some extent the rearing of poultry birds have succeeded. But the problem with the poultry owners is the absence of proper market near the village and many a time the market rates fall. So the time, money and energy spent on rearing a poultry bird for 45 days is not economically viable to him.

Duckery—This programme has severely failed and in all the groups the programme could not revive. This was mainly because of huge intake of poultry food by the Cambell ducks provided for the programmes. The villagers prefer the country ducks instead as the intake of food is less and these could survive on the available corns and millets.

Goatery—Almost every team reported a good result except in few cases some of the goats died due to the sudden disease.

Paddy Husking—The husking of paddy is one the oldest and returnable business. But in few cases it could not get due impetus due to late release of loan amount.

Eco Tourism

Three tourist resorts/hotels in the village actively participate in the tourism activity. Sundarban Tiger Camp is situated on the southeastern part of the village and is popularly known by the villagers as Wax Pol. In fact that part of the village is identified as Wax Pol. It is a luxury resort and caters to the rich and higher income groups and foreigners and perfectly blends with the surrounding lush greenery and the dense canopied mangrove forest. There are a number of tastefully decorated cottages with an eco-friendly feel in the surroundings. The resort has motorized fibreglass boats to visit some of the strategic spots of ecotourism. Very few villagers work in this resort either as a cook or boatman or as waiters.

The rest two hotels, Sundarban Green House and Sundarban Tiger View Point are comparatively a new one. Sundarban Tiger View Point is also catering to the higher income group and fast catching up with the tunes of Sunderban Tiger Camp. Comparing to them the Sundarban Green House caters to the middle-income group. Both the hotels are situated to the southwestern part of the village in between the agricultural fields and inhabited area. Near by there are two plots recently purchased for Hotel that is yet to come. In all the hotels the owners are not from the village.

The Sundarban Institute of Alternate Income Generation Sundarban mainly works in nearby villages in the Self Help Group of the eco development programme. In recent years it has also started accommodating the tourists in the institute.

Box 5 : Hotel and Resort Information

Name	*Yr. of Opening*	*Owners place of Residence/ Work*	*Permanent Worker*		*Seasonal Worker*	
			Village	*Outsider*	*Village*	*Outsider*
Sunderban Tiger Camp	2001	Kolkata		1		
Sunderban Tiger View Point	2004	Madhyam gram	2	1	5	8
Green House	2004	Gosaba	1	-	2	1

The season for tourism begins after Durga Puja in October and continues till February. From March it is almost closed. Only during Saturday, Sunday and Holidays, the tourists come in the area. The tourists come in package tour. The entry point is either Canning or Sonakhali from where the tour operators have full control over the tourists. For the Sunderban Tiger Camp and Tiger View Point, the tourists are picked up from Kolakata and brought to Dayapur directly by their own luxury bus and by boat. The tour itinerary is mainly of two types (i) one night and two days and (ii) two nights and three days. The first itinerary includes the visit to Sajnekhali Wildlife Sanctuary for view of wild life, Mangrove Interpretation Museum, Watch Tower, bonfire and folk music by the local artists, a visit to Sudhanyakhali Watch Tower and cruises through the creeks and around the cluster of islands of Sundarban Tiger Project Area. The two nights and three days itinerary includes two more added visits. One, the tour around neighbouring villages and visit to the ethnic market and second, the boat cruises inside the creeks and dense forest on way to Dobanki.

The villagers directly enter in this tourism activity in the role of folk performers or musicians who perform the cultural items of the area like Dukhe Jatra, Tushu songs and dance and some classical topics such as Mahishasur Mardini, Harishchandra-Shaiba and Nani Chor as per choice of the tour operators. Here the participation of the villager as the provision provider is negligible. The Sunderban Tiger Camp brings all the items for the tourists from Kolkata or Canning. The resort people also directly provide the emergent need of the tourists. Here the villagers work as cook or waiter or as day labour.

In case of Tiger View Point, most of the provisions come from a particular shop of Pakhirala village. Only the nearest poultry shop in the village provides the broiler items.

The case of Green House is different. It is almost in tune with the tourist lodges situated in Pakhirala village. This lodge has an understanding with one of the tourist lodge of that village who many a time pass on the tourists to them at a mutual rate due to shortage of space accommodation.

The tourists mainly come to the village for the night stay. As they come in a package tour, they are provided the lunch. Otherwise the lodge only provides the water, bedding, bathroom and tea in evening or morning. They provide food to very small group only if asked for.

Though the village has tourist resort and lodges that participate in the tourism activity but the management of the organization has been taken up by the outsiders with very few locals with getting service in these.

Impact and Peoples View

Two kinds of views are expressed. One view encourages the entry of tourism as it may boost up the economy. It may help the village in the long run. They expect some effort from the tourist lodge owners who can use the village market not only by showcasing the ethnic market but also through purchasing the items from here. More number of hotels and lodges would create more job opportunities.

The other view is diametrically opposite. The tourism business in the village is not fetching any positive note. The businessmen have set up the lodges to earn money; they are not here to help the villagers. Very few are getting jobs that also not regular as the whole tourism business is only for four months. More over, it is polluting the cultural health of the village and the young generations are affected by it and the drinking habit has increased which was mainly confined among the tribals only.

Mobile EDC bazaar in Sajnekhali and Sudhanyakhali

Box 6

Three youths of the village have been included in EDC bazaar on cooperative basis in Sajnekhali. They have started for two years and have invested Rs.7000 in it. The business is done for three months of tourist season. The forest department has given the boat for the business. The repair works of the boat is done by the cooperative members. One youth takes his mobile boat to Sudhanyakhali and sells the items there which include biscuits, hajmola, lozenges, beetle, biri, cigarette, match, chola, matar, badam, khaini, tiranga, big babool, green coconut, mineral water, honey, camera roll, battery, pen, blade, soap, vicks, crocin etc.

EDC Permanent Stall at Jetty of Sajnekhali Wild Life Sanctuary

The Pakhirala EDC formed recently one permanent stall at the jetty of Sajnekhali Wild Life Sanctuary in the line of mobile EDC bazaar. The formation of this stall has created resentment among the shopkeepers of Tiger More in Pakhirala village as well as to the members of the mobile EDC bazaar. The tourist coming to the area are purchasing the honey at a higher rate from this stall only and showing suspicion at the honey sold by others. The members of the mobile bazaar are also unable to sell their things as the tourists are choosing the stall only because it is strategically placed at the right place.

EDC Tourist Guides

Seventeen tourist guides come from neighbouring two villages, which are Pakhirala village and Jamespur village besides

Dayapur. These 17 guides worked in 2005 season of which 15 are newcomers. The recruitment of tourist guides began as early as in 1984-85 when the Sajnekhali Tourist Lodge was built in Sajnekhali National Park. During that period the Forest Department had started recruiting the villagers as daily labourers for the clearing of the forestlands, jetty ghats and other construction works. The Department selected 12 young boys as guides and absorbed them on Rs 20 per day on no work no pay basis. This continued till 1991–92 when the wage was increased to Rs. 50. But by this time 2 guides left this work and joined the family agriculture. The number of guides remained 10. One has to remember that the earning of the guides was restricted to only three months (peak) i.e. after mid October or early November to February. After that the weekends till June, could bring some tourists.

In 1997 the wage rate of the guides increased to Rs. 100. By this time the members of Bon Majdur Union, the organization of the daily labours were made permanent employees by the forest Department. But it was declared under the directives of Forest Department that no permanent job would be given to the guides, as muster roll is not maintained for them. During 1999-2000, a call was made for the recruitment of Boatman through Employment Exchange. The criterion was a certificate of experience in boat, but this was not possible for the guides, as they had no such certificate. Further when they approached any launch or boatman, they were asked to work for three months under them along with a charge of Rs. 20,000-25,000. This was not possible for the guides, so this could not help them .The recruitment for the Boatmen was completed in 2003.

In 2004 the older guides were asked to fill up the form under the EDC. But the old guides were not happy with the decision, as they did not like to be included with the juniors. They felt that they have lost the earnings. EDC has made a rule that all the earnings of a day are to be divided among the members equally.

Recently the forest department has given appointment to few youths of the fringe villages to carry out the work of tourist guides.

They terminated the old guides who were earlier working in the area and appointed the new group. Only, two of the old guides are still around and have signed up the contract. A tourist guide is paid Rs 150 per day for a tourist party.

Discussion

The issue of conservation directly leads to the issue of development in a protected area.

It is to be remembered that the village is one of the twenty-five fringe villages of Sundarban Tiger Reserve where the ecodevelopment and management practices are being carried out for a larger global issue and hence its importance lies in the role it plays. The main purpose of the ecodevelopment is to bring sustainable development that will act as a catalysing agent for the protection of the nature.

But this again brings us to the very basic issue that behind all the relationship of conservation and development lies a combination of historical, cultural and socio-political factors. While solving the issues questions are often asked that whose protected area is this? Is it helping the communities to maintain their conservation practices or undermining them by subjecting the areas into other priorities.? Is there any imposition on their will or on their traditional collection of resources? Are there mechanisms which foster equitable benefit sharing? And lastly, who is the decision maker in the development projects and issues?

Principally, the real effort is to provide long-term sustenance to the people and putting effort to local benefits. The worldwide environment consciousness has widely recognised that 'ordinary people have a central role to play in protected area management, especially those local communities reliant upon the natural resources they contain (Borrini Feyaranend et al., 2000, Kothari et al., 1996).

In the studied village of Dayapur, the declaration of biosphere resulted in loss of traditional resources. Either the resource was inaccessible or the approach route was changed and a new area was demarcated for resource exploitation. This had a greater impact on the economy and sustenance that meant a new area or the alternate avenue needed for the local population.

Two ways were open to the people. One, to search for new option on their own and second, to wait for the government programmes. People have tried for both the options. Around late 1970's, the tiger prawn seed collection could offer a good amount of earning to the local people and in 1980's it was a kind of boom with coming of aquaculture farms in parts of greater Sundarban with its network beginning from the village level. The tiger prawn collection could give employment to the men, women and children. Next to forest this was the area where easy money was readily available and women and children could supplement the earning. The unavailability of the alternate source of income in the village forced the tiger prawn seed collectors to collect the seeds. But this has caused other problems. One, the area being in high active delta zone is prone to riverbank and embankment erosion due to the turbulent river. This leads to the entry of the saline water in the settlement area and in agricultural area. The process is again aggravated due to the continuous dragging of the nets along the riverbank. What we find here is that the conservation prudence is given at the village level but it needs to be addressed from the local level to the larger market (*bheri*). It is the aquaculture owners who have the political power as well as the muscle power. A political will and awareness at the social level will have to be addressed. In a different word, the conservation process got a different kind of jerk for not providing an alternative.

Secondly, the experience has taught them that Sundarban is the nursery of fishes. The fishes of the sea enter the riverine area and estuary for fertilization and producing the eggs. After the fertilization a large part returns back to the sea but a portion moves towards the rivers. The fishermen catch fish from the different rivers, creeks, and rivulets. But due to fine netting these are destroyed. This is the reason that fishermen are not getting enough fish for their sustenance. Day by day many species are lost forever and the fishermen who are not getting enough fish feel the impact. Day by day the number of persons in tiger seed collection is getting reduced due to non-availability of seedlings.

A common villager is still unaware of the cause of protection and the legalities as a whole. To him the whole mangrove area is a vast spread of resource, which was available to him for years together and now for some reasons the forest department is not allowing them to pursue their traditional occupation. During those days the forest department used to warn them about the rough weather or it was a fair weather. These two words were very important to them. But now those days are gone. A common villager is not aware of the concept of biosphere but he is aware of the two words, the buffer area and the core area. He is not allowed to encroach the core area and some parts of buffer area. He knows that the sight of people from the forest department would bring difficult days in jail and associated penalty fines, which is dreaded by them. To them now people from department of forest are more dangerous than Royal Bengal tiger. But he is allowed to move to the open sea for catching fish. Now the question often asked by the villagers is that how they can go to the open sea without crossing or entering into the core or buffer area and also from where he would get the infrastructure to move in the deep open sea? Another zone, which is open to them, is eastern part of India east of Goral river towards Purbasha Island. But the fisherman because of the dacoits from Bangladesh avoids the alternative avenue in the form of ecodevelopment which could give opportunities in form of community development projects, individual beneficiary programmes and in tourism sector. The community development programmes brought infrastructural changes in the village, which subsequently brought changes in some areas of economic importance. The restoration of irrigation canals was a turning point in the village life. This helped in the growth of the Boro paddy cultivation as well as in the increase of pisciculture.

Though the Boro cultivation increased the scope of income generation but the markets for selling the produce are not present nearby. In their absence the farmers are unable to produce in more quantities. The season encourages the potato cultivation but there is no cold storage for keeping the potato in the nearby area. Thus though in Boro season the farmer has the scope, only the family requirement is grown. Another problem has erupted in the

Boro season. A canal carries a certain amount of water for irrigation. It is found in the village that each and every farmer tries to utilize the water as much as possible during the sowing and transplantation and at some point of time the water dries up and it is again filled up only when there is good rainfall. There are pitfalls of over utilization of the water in the beginning when there is less rain as the Boro paddy requires more fertilizers and water. Hence the crop production is restricted.

The clearing of dead water bodies like Topor khal and Malshabhanga Khal brings forth another issue. The clearing is done with the help of the villagers who are paid wages on the basis of cubic foot soil. A labour is paid on the basis of 1000 cu ft digging of soil. The main problem arises with the clearing of the soil as the villagers often complain that soil digging is not done properly and the quantity is shown much more. Many a time the contractor and the labours are not taken from the village. Here the problem arises because of the political intervention in this level. Almost every year due to several village disputes, a section of the villagers mix poison or pesticides to the water of the canal. Thus, those who invest in the fishing incur a huge loss.

The construction of brick path in the village also brings up disputes. It was found that the works carried out by the Ecodevelopment committees are the centers of the disputes and settling the scores of the political parties. For any work in the ecodevelopment committees of the village, the Ranger calls work the tender. Always a problem occurs when there is a tender call. A case may be cited to explain the situation. A problem occured centering round the construction of the brick road on the backside of Wax pol in Adibashi–Mazherpara to Sanat Mistry house. During the close of my fieldwork in February 2005, on one morning there was an argument and altercations between the villager and the members of the executive committees regarding the engagement of the labourers. The villager had a demand that the work should be given to the labourers from the village. It is a common complain among the villagers that they are not given the job and contract is given to outsiders who engage labour from the outside. The common members do not come to know about any programme of the EDC. The decision is confined within the executive members. This tender had gone to a person who is a relative to one of the executive members. On a visit to the place of construction, the author found that a power tiller was working on the road. The high school boys of the hamlet did digging work, which was done from a nearby ditch. The mud was used to raise the road. This means payment is less and the labour cost is reduced for the contractor.

The tourism-related employment for the villagers is usually the low-paying service jobs such as tour guides, porters, and food and souvenir vendors. In addition, the tourism being a seasonal occupation, they are not getting year-round employment and are laid off during the off-season. Here he is at a receiving end, he is unable to share the platform with the decision makers who is an outsider and present in the village as enterpreneur. The case of a tour guide would give an interesting insight on the plight of tourism in the village. One of the village youths who was recently recruited as a guide does not go to the Sajnekhali office most of the time, as it does not help him in earning the livelihood. And also it disturbs his tuition classes every day, which brings regular money. The guides also complain, as they have to face many problems with the tourists. Many a time they are not given treated, and are not given the right payment. They are also off loaded at Jharkhali or at Canning. They also have a charter of demands concerning the wage payment and security.

It was found in the pre pages that the local party level politics was introduced almost from the very inception of the Committee and every decision was done on the basis of party interest. The meetings have often become the place of arguments between different political groups in the village who settle their scores by differing on the issues coming in the agenda and the village development takes the second place as a result the development as a whole takes a setback. The villagers often complain that they should be made aware of all sorts of development works and associated news of the village.. The continuity of the same convenor in ecodevelopment committee who is also a member of political party in the village continuously for the three respective committees is another block. In fact his disassociation from all discussions on ecodevelopment committee and other affairs of the village with field team, was another hurdle. The representation of the women participants in executive committee was principally in confirmation with the structure of the committee but their participation in womens affairs as well as in the village is not at par. Within the executive committee they are more known as wife of so and so, the names are often forgotten. Principally year wise committee should be formed but this is never done. The annual general meeting (AGM) was due in November 2005 but till February 2006 it was not done.

Often the grievances lie that the activities of the EDC are not very transparent to the common man. The resolutions and the decisive power remain confined between the Joint Convenor, Beat Officer and one or two members of the EDC. The political intervention is another factor on which the decisions are made and many a time the executive member belonging to a different party is unaware of the development programmes. The common members most of the time remains unaware of the development activities.

The debate of development and the conservation need will continue but above all to a common villager the urgent need of the area is to develop the embankments. The tidal discharges every year during the rainy season destroy the human lives, settlement and agricultural field. The reparing of the river embankments in the dry season should be taken on an urgent basis, other wise in the words of a villager 'the whole progress of the village in the name of ecodevelopment would be washed away in one huge tidal wave and again they will have to begin from the scratch. The increase in double cropping has ensured a hope but the development will be at a standstill in the absence of the cold storage and the market where a farmer can go and deal with his produce directly and sell the items directly without the hindrance of middlemen. The self help group has encouraged the poultry in the area, but a good veterneary doctor is not always available in times of need, also in the absence of proper market nearby poultry farmer is forced to sell the produce to the nearby shops or village market at less price. The island itself has no communication within it and not well connected to the nearby town. Thus, the transportation cost increases for selling the agricultural produce in the markets. The development of communication network is also needed in case of the road as well as through the river.

Though the different sectors of the society have been involved in the conservation as well as in village level development but the effective coordination between scientists with the relevant knowledge and the biosphere managers is largely missing. Here the Anthropological knowledge is essential in understanding the local problems and issues for tackling the respective situations. Though the idea is mooted and the initiatives and appropriate institutional arrangements have been made for peoples' participation, but still it needs to be ensured that each household takes part in the decision making process. This directly leads to the empowerment of the local people in local issues whether it will be in the conservation decision or in the development decision. In words of Fisher 'meaningful decision-making also involves implementation. If a decision cannot be implemented or enforced, then the role in decision-making does not involve effective power'.(2003)

Notes

1 *Parichay o thathye Dakhin 24 Pargana Jela*

2 Interview taken from Shri Anjoli Bikash Mondol, Teacher, P.C. Sen High School on 21.01.2000

References

Annual Report, 2003-04, Sunderban Tiger Reserve, Field Director.

Borrini Feyaranend et al, 2000, *Co-management of natural resources, organizing, negotiating and learning-by-doing*. GTZ and IUCN, Kesparek Verlag, Heidelberg, Germany, 95 pp.

Bridgewater, P (not dated), *Biosphere Reserves, The Ecosystem Approach and Tourism—Finding New Paradigms in a Globalising World*. Division of Ecological Sciences.UNESCO

Fisher, R.J., 2003., Innovations, Persistence and Change: Reflections on the State of Community Forestry. In *RFCOFTC and FAO*. Community forestry: Current Innovations and Experiences. Bangkok: Regional Community Forestry Training Centre and Food and Agricultural Organization of United Nations.

Kothari et al, 1996, *People and protected areas: towards participatory conservation in India*. Sage Publications, New Delhi.

UNESCO, 1996., *Biosphere Reserves: The Seville Strategy and the Statutory Framework of the World Network*, UNESCO, Paris. 18 pp.

UNESCO., 2000, *Solving the Puzzle: The Ecosystem Approach and Biosphere Reserves*. UNESCO, Paris.

Ursula Grant, *et al*, 1998, *Biodiversity and Protected Areas: the concept and case Studies.*

Wells and Brandon, 1993, The principles and practices of buffer zones and Local participation in biodiversity conservation. Ambio.22:157-162. www.envfor.nic.in, 2005, Sunderban Tiger Reserve.

8

Evaluation of Eco-tourism Activity: A Case Study of Sundarban Tiger Reserve

Amitava Dinda

Introduction

"The target of tourism industry in 2004 of India is 3.3 million foreign tourists beside domestic tourists. But it raises a pertinent question as to whether the domestic infrastructure is sufficient enough to accommodate the said number of foreign tourists particularly in interior tourist destinations. The same concern has been shown by the Ministry of tourism, Government of India (2004). According to Joint Secretary, Ministry of Tourism, Government of India (2004), 'Infrastructure is the real problem'. However, with most tourist destinations being in the interiors, there will be enough availability of rooms" (Singh, 2004).

Eco-tourism means management of tourism and conservation of nature in a way as to maintain a fine balance between the requirements of tourism and ecology, on one hand and the needs of the local communities for jobs, new skills, income generating employment and a better status for women, on the other. The word eco-tourism is synonym of sustainable tourism. Sustainable tourism is one section of sustainable development. Sustainable development is the slogan of 21st century. Eco-tourism activity centers round tourists and involvement of people of different infrastructures such as transport and communication, hotels and restaurants, travel agents/tour operators, shops, tourist guides etc., besides maintaining ecological balance. A well managed tourism at all levels is essential for achieving successful tourism development and management. Management is necessary to coordinate and synchronize the development of different sectors, to balance competing and sometimes conflicting claims on the same limited resource base, to maximize the positive impacts of tourism development and to minimize its adverse effect. For evaluation of eco-tourism activity, the available infrastructural facilities, other aspects of tourism besides Indian National Eco-tourism Policy and Lesley France's model of strategic approaches to sustainable tourism is considered for the present study.

The developmental process of Indian National Ecotourism Policy should meet the following cardinal principles:

(i) It should involve the local community and lead to the overall economic development of the area.

(ii) It should identify the likely conflicts between resource use for tourism and the livelihood of local inhabitants and attempt to minimize such conflicts.

(iii) The type and scale of tourism development should be compatible with the environment and socio-cultural characteristics of the local community.

(iv) It should be planned as a part of the overall area development strategy, guided by an integrated land use plan and associated with commensurate expansion of public services (Bezbaruah, 2002).

Strategic approaches to sustainable tourism are discussed below in model.

Ecological Maintenance

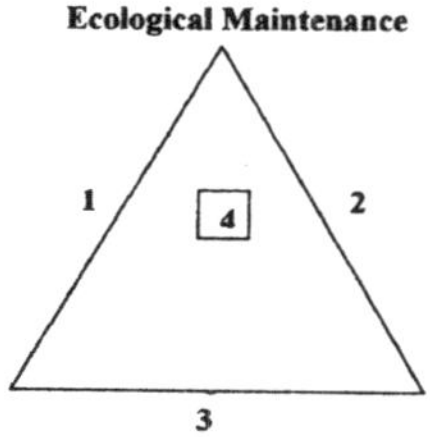

Tourist Satisfaction **Local Community**

Source: Lesley France, 1997

There are four possible strategic approaches for sustainable tourism.

Situation (1) could be demonstrated by small numbers of tourists who visit a relatively remote area, thereby gaining a high level of satisfaction from their visit and leaving their destination relatively unchanged. If such a visit is organized and operated by an external company, perhaps a multinational, benefits are unlikely to filter down to the local community and therefore will not improve the quality of life of those in most need. Specialized package holidays, like small scale safaris typically satisfy these criteria.

A small-scale local guest house could illustrate situation (2). It would provide accommodation within a physical and social environment that has been modified relatively little. The standard of comfort provided for tourists is likely to be low. Nevertheless, the original environment is preserved and any economic benefits that do accrue will go directly to the local community.

Situation (3) can occur when a large tourism enterprise employs many local people. Not all members of the local community will obtain jobs and a large enterprise such as this may well satisfy certain types of tourists but, in the process may irreparably damage the environment.

Situation (4) Small-scales, locally managed tourism enterprises that may spread benefits more widely through the community are the best example. Typical are the Casamance scheme in Senegal and ecotourism developments in Belize, where a measure of satisfaction of a variety of sustainable aims has been achieved, according to participants.

Under this backdrop of sustainable tourism, evaluation of eco-tourism activity has been studied in the Sundarban Tiger Reserve (STR), Gosaba, and its adjoining areas of 24 Parganas (South) district, West Bengal.

Land and the People

Sundarbans is the largest delta and also the largest mangrove forest in the world. The mangrove ecosystems are unique. This ecosystem is the inter-phase between sea and land, which protects the coastlines from natural calamities like soil erosion, cyclonic storms, and strong tidal water current. The Sundarbans cover an area of 10, 000, 00 ha of land and water and are part of one of the world's greatest deltas formed by sediments deposited by three great-rivers the Ganga, Brahmaputra and Meghna. Almost 62 per cent of the Sundarbans is situated in Bangladesh, while the remaining 38 per cent of the region lies in India. These two portions together comprise the largest continuous block of mangrove in the world. Indian Sundarbans are located between 21°32'-22°40' north latitude and between 88°85'-89°00' east longitudes. The Sundarban Biosphere Reserve (SBR) extends over an area of 9,630 sq. kms. in the state of West Bengal and presents a unique situation of biodiversity. The Sunderban Biosphere Reserve has three zones for coordination of its main function-Core Zone (1330 sq. kms.), Buffer Zone (1255 sq.kms.) and reserve forest of south forest division (1678 sq. kms.) which totals 4263 sq. kms. The area of Transition Zone/Non-forest area/Settlement area is 5367 sq.kms. Core zone is included in Sundarban Tiger Project as well as in the Sundarban National Park (1330 sq. kms.). The National Park area of Sundarban Tiger Reserve (STR) has also been classified as 'World Heritage Site' in 1987. Buffer Zone of mangrove forest is the area where restoration of mangrove vegetation, research and monitoring are carried out. The Transition area along the northern boundary of reserve is the reclaimed Sundarbans where agriculture and coastal aquaculture is extensively practiced. It is a dynamic zone of cooperation with local communities. This part is used for agriculture, fishing, settlements and sustainable use of local ecosystem resources (Raha, 2004).

The biosphere reserve nurtures many globally important rare and endangered species of flora and fauna. The region is home of 1100 species of Angiosperm, 150 species of Algae, 15 species of prawns, 67 species of crabs, 23 species of mollusc, 163 species of birds, 40 species of mammals, 56 species of reptiles, and 165 species of fishes. This only mangrove tiger land of the planet harbours rare and endangered mammals like-*Panthera tigris, tigris, Prionalius bengalensis, Platanista gangetica*, etc. (Debnath, 2002).

Sundarbans is the only tropical semi evergreen mangrove reserve inhabited by 'Royal Bengal Tiger' in the world. It was constituted by the Government of India in March, 1989 to protect the natural ecosystem and received the recognition of UNESCO in Man and Biosphere Programme (MAB) in November, 2001 (Raha, 2004).

Eco-tourism is practised in the Sundarban Tiger Reserve (STR), Gosaba and its adjoining region. Sajnekhali, Sudhanyakhali, Dobanki, Netidhopani and Burirdabri of Sundarban Tiger Reserve; Dayapur and Jamespur villages of Satjelia anchal; Pakhirala, Rangabelia and Jatirampur of Rangabelia anchal; Sonagaon, Arampur, Dulki, and Gosaba Bazar of Gosaba anchal, Gosaba block; Dakghat, Basanti and Sonakhali of Basanti block; and Canning station area is considered as study area.

The ethnographic atlas of the area shows an agglomeration of ethnic groups many of whom represent the tribal world. The major tribal groups of the area are the Munda, Bhumij, Oraon and the Santal. The economic pursuits of these tribal groups are based on forest economy and also food producing economy and supplementary pursuits like agriculture and agricultural labour. Even there is continuous over lapping of two or more economic activities, which have a direct bearing to their ecological conditioning and economic need. The non tribals mainly represent communities namely the Poundra Kshatriya, Bagdi, Bedia, Chamar, Jele Kaibarta, Malo, Namasudra, Rajbanshi, Brahman, Bauri, Gowala, Tanti, Kayastha, Mahishya, Napit, and so on. There are people also who belong to the faiths of Islam and Christianity, all of whom subsist on agriculture or agriculture related labour, business and other professions. There are communities who depend on forest and river for the collection of honey, wood and fish because agriculture does not fulfill their need due to saline water. The area is mono-crop in nature. The forest products like timbers, fuel woods, thatching leaves, honey and wax are the main source of commerce to local people. Besides other profession they work as cheap labour for additional source of income. During agricultural lean season, people resort to fishing by spreading net in the rivers and collection of prawn seeds even risking their lives from man-eating tigers and crocodiles. During April-May, some people also enter the Reserve Forests with permits, for collection of honey. They also collect prawn seeds from inter-dial areas. These activities cause damage to the mangrove forest and also to the ecosystem as a whole through rapid depletion of resources. Sometimes, the local people kill the tiger and sell its' skin and other body parts to businessmen in exchange of huge amount of money. The near total dependence of the people on the natural resources of the ecosystem is due to many factors. One of the reasons is uncontrolled population growth with low income level. A steep rise in population-176 per cent between 1947 and 1991 census has led to loss of forest cover in 54 of the 102 islands of Sundarbans (Debnath, 2002). The islands of Sundarbans are inaccessible even to the inhabitants and the facilities of communication like road and water transport is very poor.

Cultural Significance

The society founded upon Hindu-Muslim unity and a blend of Hindu and Muslim culture is observed in the region. Out of common fear from nature and ferocious animals the Hindus and Muslims of Sundarbans have been used to worship of Bonobibi, Dakshin Rai, Panchanan Pir, Kalu Rai, Manasa, Sitala etc. irrespective of their religious belief. These gods and goddessess are worshipped by common people generally under trees or by the riverside in open place. Bonobibi deity is worshipped by the local people on different occasions. There is no reference to them in the Shastras or Puranas. Such idols were perhaps created by the people by their own imagination to help them out of troubles. Tusu dance is common in local villages among tribals. Dukhe Yatra is also performed in local villages.

Eco-tourism Spots of Sundarban Tiger Reserve

The eco-tour spots of Sundarban Tiger Reserve are Sajnekhali, Sudhanyakhali, Dobanki, Netidhopani and Burirdabri.

Sajnekhali Mangrove Interpretation Centre-Visitors from Canning and Sonakhali first visit Sajnekhali. One mangrove Interpretation Centre, two watch towers, one crocodile pond and one turtle breeding centre are the main attractions for tourists. Deer's are easily found here and there. Bonobibi temple is also here.

Sudhanyakhali-In Sudhanyakhali one watch tower for tiger sighting and Mangrove Park is opened for tourists. Deers and wild boar are easily found here.

Dobanki-Dobanki was opened for tourists in 2003 where a canopy path running for almost 250 meters has been built 12 feet above the ground. The path is enclosed by netting that allows tourists to look down upon the dense wooded region and watch the wildlife in all its' natural splendour. Dobanki is also one of the tiger sighting zones in the STR.

Netidhopani-Netidhopani is one of the most important tiger sighting zone in the STR. One watch tower serves as tiger seeing event for curious visitors. Journey from Sajnekhali to Netidhopani is the most adventurous one. The width of the rivers is so broad here as the rivers fall in the Bay of Bengal.

Burirdabri-Burirdabri Eco-tourism complex is opened for tourists in Novembers, 2003. A shielded tunnel that runs along the ground for a kilometer into the forest is found at Burirdabri. From behind a fence, tourists can get a glimpse of wildlife in the heart of the jungle. Burirdabri is one of the 'tiger sighting zones in the STR. one watch tower is also found here for seeing wild animals. Bonobibi temple is found in all the tourist spots as well as in different points near the coastal side of the rivers.

The other tourist's interest spots of Gosaba are: Hamilton Bunglow, Bunglow used by Rabindra Nath Tagore, Gasifire (*i.e.,* electricity from wood and diesel), and Tagore Society for Rural Development, Rangabelia.

Methodology

The work was undertaken in order to have some basic and preliminary information about the present status of tourism which prevailed in Gosaba and its adjoining region. For first hand information on different aspects, various structural schedules meant for tourists, hotels and lodges, tour operators and travel agents, restaurant and tea stall owners, shop owners, tourist guides, workers involved in different sectors of tourism, and the people of different walks of life (to have their views on eco-tourism activity) were canvassed.

Despite various flaws, the schedules were randomly canvassed in different locales-Sajnekhali, Sudhanyakhali, Dobanki, Burirdabri, Netidhopani, Pakhirala, Dayapur, Gosaba, Sonakhali, Basanti, Canning and other areas. Altogether 392 samples [tourists-180 (domestic-140 and foreign-40), hotels-23, restaurants-24, tour operators-26, tourist guides-17, shops-19, people involved in different tourism sectors-44, and local people views on tourism-59] were collected on different aspects of tourism besides other information. It is admitted that the sampling made in the present study does not bear proper statistical senses. However, the same was done purposively without considering proper statistical mode because of feasibility, coverage areas, manpower, time and nature and type of subjects, etc. Besides structural schedules other standard anthropological methods are also applied for yielding field data.

Tourists, local people, and the people who are involved in different tourism sectors are asked whether they are satisfied or not satisfied about the available infrastructures and also about problems faced. For evaluation of tourism activity, the available infrastructural facility and other aspects of tourism considered for the present study are Nature Interpretation Centre, Sajnekhali (as majority of the tourists visit the spot); road condition, road and water transport, accommodation, restaurant, tour operators/ travel agents, tourist guides, shops, hospitality (as tourist flow depend on it), security (as problem of Bangladeshi and domestic poachers), cleanliness/hygiene, drinking water, telephone, electricity, and river bank condition (as flood is a regular phenomenon in the region) besides Indian National Eco-tourism Policy and Lesley France's model of strategic approaches to sustainable tourism. Intensive field work is done only in the Gosaba, and Sajnekhali area. The period of field work for collection of the present data was January and February for the years 2005 and 2006.

Aspects of Tourism

The key players in the eco-tourism business are the government, both at the central and state levels, the local authorities, the developers and the operators, the visitors and the local community. Each one of them has to be sensitive to the environment and local traditions and follow a set of guidelines for the successful development of tourism. In addition, non-governmental organizations and scientific/research institutions also have to play a key role in the development of eco-tourism. Each one of them has a set of standard operational guideline in eco-tourism (Bezbaruah, 2002). Among four major project pipelines of the Government of West Bengal, development of Sundarbans as a world class 'Eco-tourism' destination is one of the important project.

The administrative in charge of Sundarban Biosphere Reserve (SBR) is Chief Conservator of Forest (South) and Director, Biosphere Reserve, Government of West Bengal. The core area is under supervision of Field Director, Sundarban Tiger Reserve, Canning, 24 Parganas (South), West Bengal. The buffer zone is under supervision of Field Director, Sundarban Tiger Reserve (STR), Canning and also under Divisional Forest Officer, 24 Parganas (South) Division; Alipore, Kolkata. Four tourist permit issuing centers are Canning, Sonakhali, Sajnekhali and Basirhat. Recently one full time ranger was appointed for the overall in charge of eco-tourism region posted at Sajnekhali. Forest department imposes standard rules for journey and conditions for undertaking tour/ movement in Sundarban Tiger Reserve.

West Bengal Tourism Development Corporation Limited (WBTDCLtd.) is the functional agency of tourism department of Government of West Bengal. Conducted tours are organized by the Tourism Directorate as well as by the West Bengal Tourism Development Corporation. Sajnekhali tourist lodge is the only accommodation run by WBTDCLtd. in STR. Majority of tourists visit Sundarbans through privately managed tour operators/travel agents of Canning besides tour operators/travel agents of Pakhirala, Dayapur, Jamespur, and Gosaba. Limited number of tourists visit STR through government managed establishment. Tourists visit the spots by using tour operators'/travel agents' launch/mechanized boat of Pakhirala, Dayapur and Jamespur villages. Tourist spots are managed by employees of forest department.

Since the inception of Tiger Reserve, the flow of tourists had been gradually increased in Sundarban which is shown in tabular form-

Table 1: Tourists Visiting only Sundarban Tiger Reserve from 1999-2002

Year	*Domestic Tourists*	*Foreign Tourists*	*Total*
1999-2000	28,619	324	28,943
2000-2001	21,758	417	22,175
2001-2002	28,775	260	29,035

Source: Jana, 2004.

Only 0.6 per cent tourists of total tourists of West Bengal visit Sundarbans. Only 0.2 per cent foreigners of total foreigners visiting West Bengal visit Sundarbans. The flow of tourists is gradually increasing in Sundarban Tiger Reserve. In 2000-2001 the tourists flow had been decreased in STR. After that it is gradually increasing in STR (Table 1).

The flow of tourists in different entry point wise are shown in tabular form from 2002 to 2004.

Table 2: Tourist Visited STR in Different Entry Point wise from 2002-2004

Year	*No. of Tourists (per entry point) [Excluding crew]*										Total
	INDIAN					*FOREIGNER*					
	*	**	***	****	*****	*	**	***	****	*****	
2002-03	13020	7421	33824	454	54719	5	276	281	-	562	55281
2003-04	9380	14906	32909	2030	59225	-	339	297	-	636	59861

* HQ, Canning Range; ** Sonakhali; *** Sajnekhali; **** Basirhat; ***** Total

Source: Annual Report, 2003--2004, Sundarban Tiger Reserve.

Most of the tourists demand their entry permission from Sajnekhali and after Sajnekhali, Canning range, Sonakhali and Basirhat range stand later respectively. The flow of foreigners also increased year by year (Table 2). The tourist flow increased about 106 per cent in 2003--2004 compared to the period 1999-2000. The flow of tourists had been increased from the period 2002--03 to 2003--04.

On the whole 140 domestic tourists and 40 foreign tourists are interviewed during present field work. Tourists from all over the continents visit Sundarbans for enjoying the thrills of the 'mangrove-tiger' land. Among 180 tourists, 80.00 per cent tourists from Asia, followed by Europe (17.23 per cent), North America (1.67 per cent), Australia (0.55 per cent) and Africa (0.55 per cent) are interviewed (Jan-Feb, 2005). Out of 140 domestic tourists 78.57 per cent and 21.43 per cent domestic tourists are males and females respectively. Among 140 domestic tourists 89.29 per cent, 2.86 per cent, 2.15 per cent, 1.44 per cent are respectively from West Bengal, Gujarat, Uttar Pradesh and Orissa. Each 0.71 per cent domestic tourists are from Madhya Pradesh, Karnataka, Tamil Nadu, Maharashtra, Punjab and Uttaranchal. The frequency of tourists from other states is less due to inproper advertisement in media. Out of 40 foreign tourists 77.50 per cent, 10.00 per cent, 7.50 per cent, 2.50 per cent, and 2.50 per cent foreign tourists are from Europe, Asia, North America, Australia and Africa respectively. In total 57.50 per cent and 42.50 per cent foreign tourists are males and females respectively.

Eco-tourism in Biosphere Reserve promotes environmental education of the people by exhibiting nature parks and biodiversity. Awareness of natural heritage generates pride and curiosity, awe, pleasure and hopefully protectiveness towards nature. Eco-tourism plays an important role in protection of STR and conservation of natural diversity.

Due to introduction of eco-tours, the new job opportunities are opened for the local people in different tourism sectors such as hotels and lodges, restaurants and tea stalls, shops, tour operators and travel agents, tourist guides, etc. Sustainable development of local people takes place in this way.

Of all the infrastructures accommodation is most vital aspect of tourism sector. In incipient phase of tourism in Gosaba region availability of hotels and lodges were very scanty. But with the flow of tourist traffic in due course its necessity was badly felt and during 1990s five hotels came up in the Gosaba and its adjoining areas. The total number of hotels and lodges found in the Gosaba and its adjoining area is 23. Out of 23 hotels and lodges, in Pakhirala village 8 private hotels and one Zillah Parishad Guest House (39.13 per cent of total accommodation), in Dayapur village 3 private hotels and one non-government organization acting as accommodation sector in tourist seasons (17.39 per cent), in Sajnekhali Government Tourist Lodge (4.35 per cent), in Gosaba Bazar 6 private hotels (26.09 per cent) and in Sonakhali and Basanti area 3 private hotels (13.04 per cent) of total accommodation sector do their business. Sundarban Tiger Camp hotel of Dayapur collects both domestic and foreign tourists. They have their own mechanized boat. Of total 16 permanent employees, 12 employees are from adjacent villages and four are outsiders. Sundarban Tiger Camp hotel and Sundarban Tiger View Point hotel of Dayapur village act as a large tourism enterprise and are following Lesley France's situation number (3) in Indian scenario. WBTDC Ltd.'s tour is also following Frances' situations number (1). But Hotel

Madhuban, Hotel Krishna Kunja, Hotel Aram, Hotel Purbasha, Promila Tourist lodges, Amari Lodge, Sundarban Green House and all the hotels of Gosaba Bazar are following Lesley Frances' situation number (4). Affluent tourists generally avail the facilities of Sundarban Tiger Camp and Sundarban Tiger View Point hotels of Dayapur village and also avail the tour conducted by WBTDC Ltd.'s by M.V. Madhukar, M.V. Chhitrarekha and M.V. Sarbajaya along with the facility of Sajnekhali Tourist Lodge. Budget travellers generally stay in hotel Madhuban, Hotel Aram, Purbasha Lodge, Promila Tourist Lodge, Hotel Mangrove, Hotel Krishna Kunja, Swastik Lodge of Pakhirala village and Sundarban Green House of Dayapur village and all the hotels of Gosaba Bazar area. Out of 23 hotels and lodges, six hotel owners are from Kolkata. Due to the introduction of new hotels, the land price of Pakhirala and Dayapur villages is increasing day by day.

Hotel and Lodge owners and employees of hotels and lodges are the important tourist managers. In total 23 hotel owners families, 80 families of employees are directly dependent on tourism business. Indirectly the supplier of goods *i.e.,* local villagers are also benefited through tourism. Most of the hotel employees are from villages of Pakhirala, Dayapur, Jamespur, Dulki, Sonagaon, Rangabelia, Arampur, Kachukhali, Bijoynagar, Manmathanagar and Kumirmarhi of Gosaba block. Very few employees are from outside of Gosaba block in hotel sector.

The availability of rooms and other infrastructural facilities are very important factors in interior tourist destinations. From the field data, it is found that the total number of rooms available for tourist accommodation in Pakhirala, Dayapur and Gosaba Bazar is about 170. About 625 tourists can be accommodated in 170 rooms (February, 2006). In peak season there are about 20-25 launches from Canning which gather in Sajnekhali. Each launch carries about 40-45 tourists. About 800 to 1100 tourists generally arrive in Sajnekhali each day by launch from Canning. On Christmas day and on New Years Eve, the flow of tourists is high. On these days the tourists also occupy the villagers' residence as the rooms are not available besides tourists' stay in launches. Hotel/lodge owners arrange Dukhe Yatra, Tusu dance, Mahishasurmardini, folk song, folk dance, etc., for tourists in the evening. Performers are from local villages.

All the launches from Canning provide food facility to tourists. Every hotel provides food to tourists. Out of 24 establishments, 13 tea stalls and 11 restaurants act as restaurants. In such an interior destination 24 restaurants and tea stall owner's families are directly benefited through eco-tourism.

Tour operator and travel agent are the most important constituents of tourism. The success of tourism in particular spot depends on the works of tour operators/travel agents. Overall, 26 tour operator/travel agents are surveyed during the field work. Tour operators/travel agents business require three to ten persons. There are about 50 tour operators/travel agents at Canning whereas there are about nine tour operators/travel agents in Gosaba. Each tour operator/travel agent having average 7 employees manages their business. Employees of tour operators travel agents are from Pakhirala, Dayapur, Gosaba Bazar, Masjidbati, Canning, Basanti, Diamond Harbour, and Sonarpur of 24 Parganas (South) district; Hasnabad, Hingalgunj and Basirhat of 24 Parganas (North) district; and Geokhali of Purba Medinipur district. Sometimes tour operators/travel agents also directly arrange local cultural programmes in hotels for tourists.

Tourist guides are the cultural ambassadors of the country. Without tourist guide, tourism can not sustain. Biodiversity of Sundarbans can not be understood without tourist guide facility. Forest department provides tourist guide in each boat or streamer. Forest department recruits 23 tourist guides from Eco-Development Committees (EDC) of Pakhirala, Dayapur, and Jamespur villages of Gosaba block. Out of 23 tourist guides only 17 tourist guides perform their job (Feb, 2006). In total 17 tourist guides are interviewed.

Shops are necessary for tourism. In total 19 shops are studied during field work. Petty gift shop, telephone booth, petty book shop etc., are found in Sajnekhali, Pakhirala and Dayapur village, Gosaba Bazar and Sonakhali. The items sold by shops are honey; tender coconut; mineral water; replica of tiger, deer, crocodile, shark, crab and Bonobibi deity; etc. Some of the shop owners also sell Sundari and Kakra tree, Kath Badam, Bee hives, etc.

People of different age groups, castes involve in different tourism sectors. People involved in different tourism sectors are from different villages of Gosaba block, Basanti block, Canning, and Diamond Harbour of 24 Parganas (South) district; Hingalgunj, Hasnabad, and Basirhat of 24 Parganas (North) district; Kantai and Geokhali of Purba Medinipur district. Among different workers, only 44 workers are interviewed. Out of nine ladies, seven as hotel workers and two as restaurant workers perform their duty in Gosaba. Caste-wise classification of 44 workers Poundra Kshatriya (65.92 per cent), Mahishya (20.45 per cent), Dhopa (4.54 per cent), Tantubai (4.54 per cent), Jele Kaibarta (2.28 per cent), and Muslim (2.27 per cent) perform duties as hotel workers, restaurant workers, travel agent workers, van rickshaw pullers etc.

The fruits of economy flourishing on account of tourism are not percolating down to the villagers of Pakhirala, Dayapur, Jamespur, Gosaba market area and its adjoining areas as is evident from their economic condition due to their lesser role in tourism. Besides this, the other contributing factors like-ignorance, inaccessibility, poor economy etc., are responsible for their lesser role in tourism.

A good number of people are involved in managing tourism in Canning Station area. They are involved in supplying vegetables, fish, meat, drinking waters grocery items, egg, generator, etc. to the travel agents. A good number of people are involved in marketing and advertisement of eco-tourism of Sundarbans in Canning station and also in Kolkata.

For evaluation of eco-tourism activity on the available infrastructural facility and other aspects of tourism 392 informants *i.e.*, tourists, local people, and the people who are involved in different tourism sectors are interviewed whether they are satisfied or not satisfied about the available infrastructures.

Tourists, People involved in different infrastructures and local people's views on available infrastructural facility and other related aspects of tourism are given below in tabular form—

Table 3: Tourists, People involved in different infrastructures and local people's views on available infrastructural facility and other related aspects of tourism

N = 392

Sl.No.	Category	Satisfactory		Not Satisfactory		No Comments		Total	
		No	%	No	%	No	%	No	%
1.	Nature Interpretation Center (Sajnekhali)	284	72.45	78	19.90	30	7.65	392	100
2.	Road condition	135	34.44	231	58.93	26	6.63	392	100
3.	Road and Water Transport	243	61.99	118	30.10	31	7.91	392	100
4.	Accommodation	259	66.07	72	18.37	61	15.56	392	100
5.	Restaurant	281	71.68	72	18.37	39	9.95	392	100
6.	Tour operator/ travel agent	253	64.54	69	17.60	70	17.86	392	100
7.	Tourist Guide	176	44.90	144	36.73	72	18.37	392	100
8.	Shop	233	59.44	118	30.10	41	10.46	392	100
9.	Hospitality	344	87.75	11	2.80	37	9.45	392	100
10.	Security	224	57.14	128	32.65	40	10.20	392	100
11.	Cleanliness/ Hygiene	339	86.48	18	4.59	35	8.93	392	100
12.	Drinking Water	28	7.14	320	81.64	44	11.22	392	100
13.	Telephone	12	3.06	332	84.70	48	12.24	392	100
14.	Electricity	12	3.06	332	84.70	48	12.24	392	100
15.	River bank Condition	20	5.10	331	84.44	41	10.46	392	100

Majority of tourists, hotel and lodge owners, restaurant and tea stall owners, tour operators/travel agent owners, tourist guides, shop owners, worker involved in different tourism sectors, local people views on tourism activity are not satisfied in the following infrastructural fields such as drinking water (81.64 per cent), telephone (84.70 per cent), electricity (84.70 per cent), river bank condition (84.44 per cent), and road condition (58.93 per cent). Informants are not satisfied in the following fields-road and water transport (30.10 per cent), tourist guide (36.73 per cent), security (32.65 per cent), and shop (30.10 per cent) [Table 3].

Bar chart showing frequency distribution of tourists, people involved in different infrastructures and local people's views on available infrastructure and other aspects of tourism

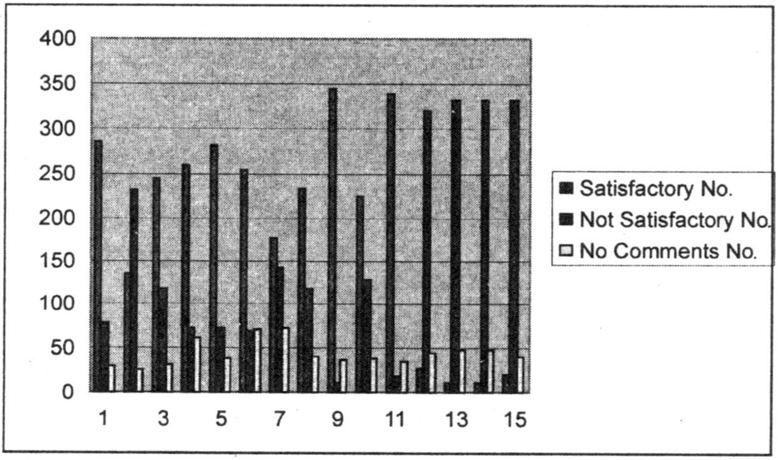

The major problems faced by tourists, hotel, restaurant, travel agent, and shop owners, tourist guides, workers, and local peoples are given below in tabular form-

Table 4: Major problems faced by tourists and others

Sl. No.	*Category*	*Major problems faced*
1.	Tourists	Proper infrastructural facilities at entry points; drinking water, waiting hall, proper toilet, more number of jetty ghats in all tourist spots; etc.
2.	Hotel owners	Tourist season only three months, Advertisement in print and electronic media about the tourist spots, better transport, more in number and better tourist spots, etc.
3.	Restaurant owners	Only three month tourist season, need for proper advertisement in media, etc.

4.	Travel agent owners	Proper infrastructures in entry points as well as tourist spots, more number of jetty ghats on tourist spots, shallowness of river bed, fishing nets in rivers, tourist guide's off loaded problem, itinerary of visits, tour and foreigners entry permission from Kolkata, some of the tourist's drunkenness, launch and boats entry fees high, etc.
5.	Tourist Guides	Tourist season only three months, Proper advertisement in media, off loaded problem after tour, communication in English, Hindi and other foreign languages, waiting hall, etc.
6.	Shop owners	Only three months business, proper advertisement in media, etc.
7.	Workers	Low payment, duty hours so long, only three months tourist season, etc.
8.	Local Peoples	Water and road transport, drinking and irrigation water, flood, garbage and price hike of commodities at Pakhirala, land price hike at Pakhirala and Dayapur villages, tourist's dress, free mixing, etc.

Positive and negative impact of tourism is given below:

Impact of tourism on local villages

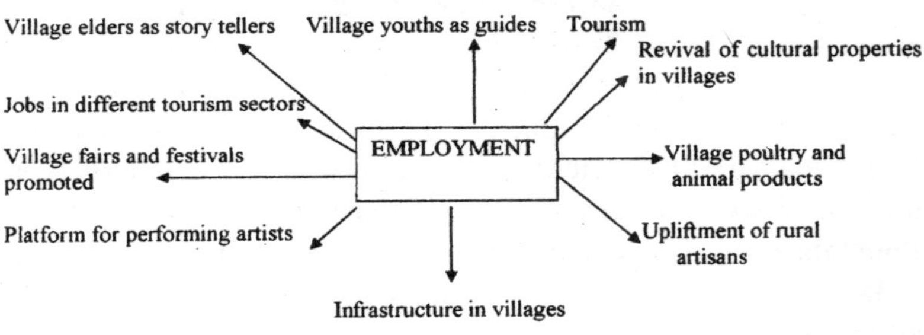

These are the spheres, where tourism can play an important role.

The persons who have vast experience about the jungle (especially those were attacked by tigers) of Pakhirala and Dayapur villages are Bauley (those who collect wood from jungle) and Maulay (who collect honey from jungle and having the knowledge about mantras of tigers' movements in the jungle) and Jele (who catch fish) of Pakhirala and Dayapur villages, sometimes the priest of Bonobibi Puja and others are the main attraction for the tourists as story tellers. Dukhe Yatra is one of the important cultural performances which get rejuvenated due to tourism. Baba Tarak Nath Opera of Dayapur village perform Dukhe Yatra, Mahishasurmardini, Krishnalila, Tusu dance and other cultural programmes in different hotels and lodges of Pakhirala and Dayapur villages. Employees of Sundarban Tiger Camp perform Dukhe Yatra in their hotel. For Dukhe Yatra about twenty persons are needed. In Dukhe Yatra local groups demand about fifteen hundred rupees. Besides this amount, the local groups also collect tips (in rupees) during their performance such as '*bhikha de ma nagarbasi*' in Dukhe Yatra. Local artists including two boys and girls also perform Baul and Adhunik song, Bratachari nritya, Rabindra sangit, etc. Both the tribals and non-tribals perform Tusu dance in different hotels. Mahishasurmardini, Krishnalila,Tusu dance and other cultural programmes need about three to ten persons. In Mahishasurmardini, Krishnalila, Tusu dance and other cultural programmes, local groups demand about five to six hundred rupees besides tips (in rupees). Tourists are very much impressed by observing such enriched local culture. By interacting with the local artists, tourists also perform songs, music, etc. Bonobibi, Durga, Saraswati and other pujas are in the tour programmes of different hotels. During Durga puja, tourists also distribute clothes to local boys, girls and others. Artisans both tribals and non tribals of Dayapur and Pakhirala villages are now busy in making replica of tiger, deer, crocodile, crab, Bonobibi deity, etc., which are the sale items to the tourists. Van rickshaw pullers of Pakhirala, Rangabelia, Dulki, Arampur and adjoining areas earn major income from tourists in peak tourist season.

In addition to economic development, intercultural contact takes place through conversation, and cultural programmes participated by both the local people and tourists. When tourists and local people come together, both have the opportunity not only to know how they live, but also to reflect on their own lives through the eyes of others. Dress, use of cosmetics, liquor consumption, etc., are frequently borrowed by the local people from tourists. As a result cross-cultural interactions occur. Awareness of fundamental rights, education, availing of better health facilities etc., enrich them while making contact with the domestic as well as foreign tourists.

Frequent movements of launches and *bhatbhati* (locally made boat) break the solitude of Sundarban Tiger Reserve which hampers the normal life of wild animals and plants besides oil spillage in the rivers.

Local villagers opine that tourists' drunkenness, dress, free mixing ('*drishya dushan*' termed by local people), etc, are not healthy for villagers.

Discussion

Sundarban Tiger Reserve's unique 'mangrove-tiger' land attracts eco-tourists though in limited number from all over the globe. Since the inception of the Tiger Reserve the tourist flow in Sundarban Tiger Reserve has been sharply increased to about 106 per cent in the period 2003-2004 as compared to the period 1999-2000. Gosaba *bazaar* (market), Pakhirala and Dayapur of Gosaba block are the fringe villages of Sundarban Biosphere Reserve and tour initiating centres to Sundarban Tiger Reserve (STR) and also point of tourist attraction and tourist halt. Over the years a steady flow of both domestic and foreign tourists from all over the continent have registered at this place. However, the fruits of economy flourishing on account of tourism are not percolating down to the villagers of Pakhirala, Dayapur, Jamespur, Gosaba market area and its adjoining areas as is evident from their economic condition due to their lesser role in tourism. Coupled with it, the other contributing factors are: ignorance, inaccessibility, poor economy, etc. Only recently the local people have begun participating in small scale private entrepreneurs like-hotels, travel agents, shop owners, etc.

It is observed that most of the eco-tourists visit the Sundarban Tiger Reserve by availing of the facilities of Canning tour operators/ travel agents. So, involvement of people from Canning and its adjoining areas is much more than the people involved in eco-tourism of Pakhirala, Dayapur, Jamespur and its adjoining areas of Gosaba block. Hence, it does not lead to the overall economic development of Pakhirala, Dayapur, Jamespur and its adjoining areas of Gosaba block. It does not follow Indian National Eco-tourism Policy.

Due to mono-crop area and high influx of population in adjacent villages of Sundarban Tiger Reserve, local villagers are forced to depend on forest based resources. Local villagers illegally fell mangrove trees and sell its timbers and branches to businessmen. Sometimes poachers kill the 'Royal Bengal Tiger' and sell its skin and other body parts to businessmen in exchange of huge amount of rupees. Local villagers also depend on fishing by spreading nets in the rivers. The eco-tourist's streamers also travel through the same route. Sometimes they result in conflict. Hence, it is observed that the local authorities somehow can not follow the guidelines of Indian National Eco-tourism Policy.

Out of the 392 informants, majority of tourists, hotel and lodge owners, restaurant and tea stall owners, tour operators/ travel agent owners, tourist guides, shop owners, workers involved in different tourism sectors, local peoples' views on tourism activity are not satisfied in the following infrastructural fields such as drinking water (81.64 per cent), telephone (84.70 per cent), electricity (84.70 per cent), river bank condition (84.44 per cent), and road condition (58.93 per cent).

It is suggested that improvement in a few sector can help in flourishing eco-tourism in Sundarban Tiger Reserve:

- Improved water and road transport, better road, over bridges (Canning-Dakghat, Gadkhali-Gosaba), availability of electricity, telephone facilities, etc.

- Number of jetty ghat may be increased in all tourist spots as well as Pakhirala Tiger Morh and Dayapur.
- Facilities of Drinking water, waiting hall/resting place, proper toilet facility, etc., may be made available in all tourist spots as well as at all entry points.
- Shallow river bed, setting fishing net in the rivers, problem of taking tour and foreigners entry permission from Kolkata, tourist guides' off-loaded problem after tour, drunkenness, mood of some of tourists, etc., are the major problems for the tour operators/travel agents.
- The major problems faced by the workers involved in tourism sectors are low payment and prolonged duty hours.
- Problem of communication skills in English, Hindi and other foreign languages, off loaded problem after tour, waiting hall at Sajnekhali, etc. are the common problems of tourist guides of eco-development committee of local villages.
- The basic needs of the local people, i.e, water scarcity both drinking and irrigation should be developed on priority basis.
- River bank should be protected from flood by any means.
- Dukhe/Bonobibi yatra, Tusu and Folk dance, Folk songs, etc., are the sources of income some of local villagers in tourist season. It is observed that Government aid is needed to encourage them in flourishing their folk culture.

Jobs in different tourism sectors, village poultry and animal products, village youths as guides, village elders as story tellers (about tiger, fishing and honey collection), platform for performing artists (dukhe yatra, tusu dance, folk dance, folk song), rural artisans, etc., are the products of eco-tourism of Sundarban Tiger Reserve.

Frequent movements of launches and locally made mechanized boats break the solitude of Sundarban Tiger Reserve which hampers the normal life of wild animals and plants besides oil spillage in the rivers.

It should be incorporated village are not congenied to the local culture. Garbage problem and price hike in tourist seasons at Pakhirala Tiger Morh create problem to villagers. Land price hike at Pakhirala and Dayapur villages is problem for villagers.

Eco-tourism acts as a protective agent of Sundarban Tiger Reserve. 'Mangrove-Tiger' land should be protected in order to prevent natural calamity in deltaic West Bengal as well as for eco-tourism purpose.

Eco-tourist's degree of satisfaction depends on several factors. Some of the important factors are availability of proper infrastructural facility, involvement of local community in ecotourism activity, proper tourism management, maintenance of eco-tour spots and fragile ecosystem.

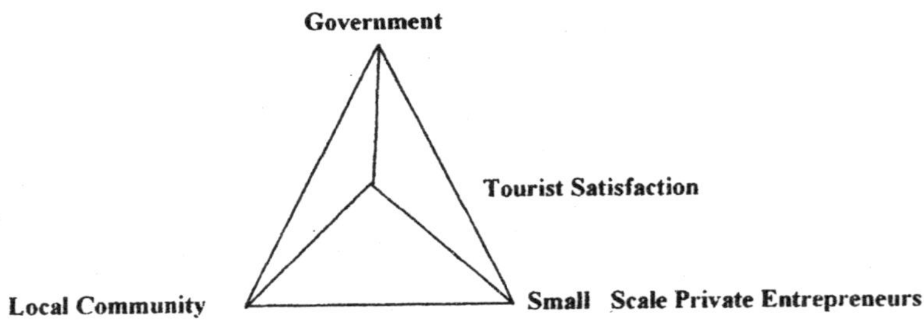

Strategic approach to sustainable tourism for Sundarban Tiger Reserve

In the management process of tourism, the interplay of three indicators *viz.* local community, small scale private entrepreneurs and government determine the degree of management in terms of availability of infrastructure, conservation of ecology, benefit to local community and satisfaction to tourists. In case of Sundarban Tiger Reserve, which has a fragile ecosystem, a greater participation of local community along with decisive government intervention in management of tourism is suggested to safeguard the interest of the local community and protect the ecology. It, in turn, will ensure maximum satisfaction to tourists. As far as private entrepreneurship is concerned, the small scale private entrepreneurships' involvement should be encouraged.

If government and small scale private enterprises act together properly it may spread benefits more widely through the community, there would be chance of huge employment opportunities to local villagers of Pakhirala, Dayapur, Jamespur and its' adjoining villages of Gosaba block and villagers of Basanti block. It is earnestly suggested that the Forest Department and other concerned departments should take initiative for involving local villagers in eco-tourism. The Forest department's role should be highly sensitive and they should act more like professionals in a cooperative way.

Increasing numbers of visitors to ecologically sensitive areas like Sundarban Tiger Reserve can lead to significant environmental degradation. Likewise, indigenous culture faces numerous challenges with the influx of foreign visitors and wealth. Eco-tourism promotion is bound to have a series of economic, socio-cultural, environmental impacts on the habitats which would require a systematic evaluation and analysis using scientific techniques before development decisions are taken.

A well coordinated managerial skill on different infrastructures of tourism involving local community as well as maintenance of fragile ecology could not sustain against the subsistence pattern of local community that leads to the highest degree of tourist satisfaction. Hence, it is suggested that management of tourism infrastructures and tourism resources should be done involving local community in such a way in Sundarban Tiger Reserve that it fulfills economic and social needs of the local community while maintaining cultural integrity, essential fragile ecological processes, biological diversity and maximum satisfaction to tourists.

References

Bezbaruah, M. P., 2002, *Frontiers of New Tourism*. Gyan Publishing House, New Delhi.

Debnath, H. S., 2002, *Sundarban Biosphere Reserve*. Botanical Survey of India, Kolkata.

Field Director, 2004, *Annual Report, 2003-2004 on Sundarban Tiger Reserve*. Sundarban Tiger Reserve, Canning, W.B.

France, Lesley ed, 1997, *The Earthscan Reader in Sustainable Tourism*. Earthscan Publications, U. K.

Jana, Debaprasad, 2004, *Sree Khanda Sundarban*. Deep Prakasan, Kolkata.

Raha, A. K. ed, 2004, *Indian Sundarban-An overview*; Director, Sundarban Reserve, Forest, Wildlife Wing, Forest Department; Kolkata.

Singh, Garima, 2004, Infrastructure key to tourism boom. *The Telegraph*, 21st September.

List of Contributors

1. **D. B. Mandal**, Deputy Director and Head of Office, Eastern Regional Centre Anthropological Survey of India, EN-79, Sector-V, Salt Lake City, Kolkata-700 091.
2. **Sudhansu Gangopadhyay**, Cartographer, Eastern Regional Centre, Anthropological Survey of India, EN-79, Sector-V, Salt Lake City, Kolkata-700 091.
3. **Archna Datta**, [Fortner Research Associate (Human ecology) Anthropological Survey of India, Kolkata], 89/144, Bangur Park, Post-Rishra, Dist. Hooghly-712248, e-mail-datta archna@yahoo.com
4. **Kakli Chakrabarty**, Superintending Anthropologist (Cultural), Eastern Regional Centre, Anthropological Survey of India, EN-79, Sector-V, Salt Lake City Kolkata-700 091.
5. **Krishna Basu**, Research Associate (Cultural), Eastern Regional Centre, Anthropological Survey of India, EN-79, Sector-V, Salt Lake City, Kolkata-700 091.
6. **S. S. Mishra**, Assistant Anthropologist (Cultural), Eastern Regional Centre Anthropological Survey of India, EN-79, Sector-V, Salt Lake City, Kolkata-700 091.
7. **Krishna Mandal**, Anthropologist (Cultural), Eastern Regional Centre, Anthropological Survey of India, EN-79, Sector-V, Salt Lake City, Kolkata-700 091.
8. **Subrata Ray**, Research Associate (Cultural), Head of Office, Anthropological Survey of India, 27, Jawaharlal Nehru Road, Kolkata 700 016.

9. **Ratna Dhar**, Anthropologist (Cultural), Head of Office, Anthropological Survey of India, 27, Jawaharlal Nehru Road, Kolkata-700 016.

10. **Amitava Dida**, Research Associate (Cultural), Eastern Regional Centre, Anthropological Survey of India, EN-79, Sector-V, Salt Lake City, Kolkata-700 091.

Index

F

M

N

T